S0-AXJ-132

A
Programmed
Review For
Electrical
Engineering

A Programmed Review For Electrical Engineering

Second Edition

JAMES H. BENTLEY, P.E.
Magnetic Peripherals, Inc.
Minneapolis, Minnesota

Programmed by
KAREN M. HESS, Ph.D.
Innovative Programming Systems, Inc.
Minneapolis, Minnesota

VNR VAN NOSTRAND REINHOLD COMPANY

Library of Congress Catalog Card Number: 83-21835
ISBN: 0-442-21628-9

Manufactured in the United States of America

Published by Van Nostrand Reinhold Company Inc.
135 West 50th Street
New York, New York 10020

Van Nostrand Reinhold Company Limited
Molly Millars Lane
Wokingham, Berkshire RG11 2PY, England

Van Nostrand Reinhold
480 Latrobe Street
Melbourne, Victoria 3000, Australia

Macmillan of Canada
Division of Gage Publishing Limited
164 Commander Boulevard
Agincourt, Ontario MIS 3C7, Canada

15 14 13 12 11 10 9 8 7 6 5 4 3 2 1

Library of Congress Cataloging in Publication Data

Bentley, James H.
  A programmed review for electrical engineering.

  Includes index.
  1. Electrical engineering—Problems, exercises, etc.
2. Electric engineering—Programmed instruction.
I. Hess, Karen M., 1939–0000. II. Title.
TK168.B46 1984   621.3'07'7   83-21835
ISBN 0-442-21628-9

Preface

The field of electrical engineering is very innovative—new products and new ideas are continually being developed. Yet all these innovations are based on the fundamental principles of electrical engineering: Ohm's law, Kirchhoff's laws, feedback control, waveforms, capacitance, resistance, inductance, electricity, magnetism, current, voltage, power, energy. It is these basic fundamentals which are tested for in the Professional Engineering Examination (PE Exam).

This text provides an organized review of the basic electrical engineering fundamentals. It is an outgrowth of an electrical engineering refresher course taught by the author to candidates preparing for the Professional Engineering Examination—a course which has enabled scores of electrical engineers in Minnesota and Wisconsin to successfully pass the PE Exam.

The material is representative of the type of questions appearing in the PE Exams prepared by the National Council of Engineering Examiners (NCEE) over the past twelve years. Each problem in the text has been carefully selected to illustrate a specific concept. Included with each problem is at least one solution. Although the solutions have been carefully checked, both by the author and by students, there may be differences of interpretation. Also, in some cases certain assumptions may need to be made prior to problem solution, and since these assumptions will vary from individual to individual, the final answer may also differ. The author has attempted to keep the requirements for assumptions and interpretation to a minimum. However, since problems requiring this sort of judgment appear in the PE Exam, they have also been included in this text.

Although the types of questions are similar to those given in past exams, the author cautions that the NCEE may include questions in future exams that may differ considerably from past format. Each exam is new, with very little repetition from past exams. However, the purpose of the exam is to test the examinee's knowledge of electrical engineering fundamentals, and if the examinee, through experience, education, and review, has command of these fundamentals, he will have no trouble in passing the exam.

While the material in this text is intended as a programmed review for those preparing for the PE Exam, the material is fundamental to electrical engineering and can provide an excellent resource and review for any student or practicing electrical engineer. Rarely does an electrical engineer work with all areas of electrical engineering on a daily basis. However, during the course of an engineer's career, he will probably become involved in nearly every field. (This has been the author's experience, and the material in this text has proved invaluable on numerous occasions.)

The author has conducted refresher courses since 1973 using the material contained in the text. Students have come from such diverse enterprises as 3M Company, General Electric, IBM, Northern States Power, Univac, Medtronic, Honeywell, Northwestern Bell, the State of Minnesota, Control Data, and a number of private consulting engineering firms. The author wishes to thank these students for their valuable suggestions, criticisms, and corrections. Others to whom the author is deeply indebted include Allen Baldwin, Connie Nelson, Fred Pfeiffer, and Francis Cook.

James H. Bentley, P.E.
Minneapolis, Minnesota

Introduction

This text reviews electrical engineering fundamentals required to pass the Professional Engineering Examination (PE Examination) in Electrical Engineering. It is designed to aid the electrical engineer in preparation for the PE Examination by presenting an organized review of materials ordinarily contained in a college electrical engineering curriculum. The primary emphasis is on problem solving with orientation as close as possible to the type of questions contained in the exam.

WHY REGISTRATION?

The requirement for engineering registration is continually spreading throughout the United States. In 1975, 40% of U.S. engineers (nearly 500,000) were registered. Most engineers in private practice or working for state or local agencies must be licensed. Many engineers in industry need to be registered to sell their designs and products to certain consumers. While most states presently exempt engineers in industry from the requirement for registration, the trend is toward removal of this exemption. As noted by Fran Lavoie, managing editor of *Machine Design* (December 26, 1974): "If you hope to rise to a responsible engineering position, you'll eventually have to be registered. Conversely, the fact that you aren't registered could cost you a promotion."

Requirements

There are six basic requirements for registration (the specific requirements may vary slightly from state to state): (1) *age*—minimum age of 25 for licensing, 21 for Engineer-in-Training; (2) *graduation* from an accredited high school; (3) an *engineering degree* from an accredited ECPD course of study in an engineering school or its equivalent in practical engineering experience; (4) *experience* of a minimum number of years, broad in scope and generally requiring knowledge of engineering mathematics, physical and applied sciences, properties of materials, and the fundamental principles of engineering design; (5) *character* references; and (6) *examination* by the state board of engineers. Some states also require U.S. citizenship.

You should contact your licensing board to obtain the specific requirements for your state or jurisdiction. The mailing adresses of Member and Affiliate Member Boards may be obtained from the National Council of Engineering Examiners, P.O. Box 5000, Seneca, SC 29678.

A LOOK AHEAD TO THE PE EXAM

The PE Exam tests an individual's experience and knowledge in his field; it is intended to show the applicant's ability to apply sound engineering principles and judgment to the solution of problems encountered in practice.

The exam is divided into two four-hour periods. You must work four problems during each period: four engineering during one period, and one economics and three engineering during the other period. (In some states the economics problem is optional and may be replaced by a fourth engineering problem.) During each four-hour period there are ten engineering problems from which to select. The major categories which may be covered during the eight-hour examination are:

- power and systems
- machines
- electronics
- communications
- circuits
- controls
- economics
- instrumentation
- illumination
- logic

The PE Exam is straightforward. There are not trick questions. You are either familiar with a particular problem or you are not. Therefore, before working any engineering problems, read them *all* and eliminate those with which you are least familiar. Select the problems that best fit your background and work them. (You have an average of one hour per problem.) Do the easiest problems first; concentrate the remainder of your time on the more difficult ones. Try not to make any computational errors. However, if you do make an error, partial credit is given if your basic approach is correct. Therefore, it is important to show your method of solution as well as all your work.

PREPARATION

Like anything done well, taking the PE Exam requires thorough preparation. Seldom can a person walk into the exam with little or no preparation and expect to pass.

To do well on the exam, allocate yourself about ten weeks of review time. Read a chapter from this text each week. Work all the problems. Then rework them. Refer to your college texts for other problems and background material. Allocate the week before the exam for a general review. You may concentrate your efforts in areas you are most familiar, but do not expect the PE Exam to include questions in every area of electrical engineering. It is best to be familiar with as many areas as possible. One time the PE Exam may have several problems on machinery or control theory, and six months later there may be none.

What to Take Along

Most states administer an open book examination. Your state board will let you know what you can use during your examination. Some examinees take a suitcase or two of reference books, but usually only one or two of these books is actually used. Some problems require use of handbooks, but general textbooks usually are not needed.

The following is a recommended list of basic items to take to the exam (if permitted by your board of registration):

- *CRC Standard Mathematical Tables*
- *Reference Data for Radio Engineers*
- a general electrical engineering handbook
- an electrical engineering review book (this text might be used)
- a calculator (fully charged)
- compass and straight edge (for Smith chart problems)
- six sharp pencils with good erasers

A SYSTEMATIC, PROGRAMMED REVIEW

This text is organized to provide a logically sequenced review of electrical engineering and economics. It is designed around problems and provides the necessary information needed to solve the problems in the form of tables, formulas, charts, and graphs. It does *not* contain background material that can be found in other texts and handbooks. Where background material is desired, the reader is directed to other source material. An extensive bibliography is also included. The serious reviewer will refer to other texts and handbooks for derivations of formulas and other background materials needed for a complete review.

The first chapter of the text introduces the basic principles of electrical engineering. The second chapter is devoted entirely to problems related to these fundamental principles. It is intended as an applied review of basic circuits and the concepts presented in the first chapter. The remaining chapters cover the several specific areas of electrical engineering where mathematical and graphical techniques supplement the fundamental concepts. In these chapters a discussion of background materials needed for solving problems is normally given. The last chapter deals with economics, a subject representing $\frac{1}{8}$ of most PE Examinations.

The book is based on problems, just as is the PE Exam. Discussion is included to introduce a subject and to provide the background necessary to solve a particular problem. In some instances a general discussion precedes one or more problems, in which case the problems refer back to the discussion. In other cases, a problem will be self-contained; the solution given for the problem will contain all the needed information; no reference will be made to preceding discussions.

The text attempts to include problems that, in toto, illustrate nearly every concept and every type of question asked on the NCEE-prepared PE Exam during the past eighteen years. In some cases the problems included are subsets of past PE Exam problems, broken down to illustrate specific points. A PE Exam problem can comprise the contents of several of the problems presented in this text. There may also be variations on a particular problem. No one but the NCEE knows what questions will be on the next PE Exam until the time of the exam.

This text omits two areas sometimes included in the PE Exam—illumination and national electric code—and it includes one area covered very seldom in the PE Exam—logic. Illumination is not covered here because very few engineers are involved in the field, it is not included in traditional electrical engineering curricula, and it is based more on aesthetics and empirical data than on fundmentals. Similarly, national electric code problems are not included because they apply only to a limited number of engineers. Examinees working in either of these fields will have the necessary experience and handbooks to solve the problems. Others may simply skip these problems.

Logic is covered because a significant number of electrical engineers use digital techniques in their work. Logic touches the lives of nearly everyone and has an important impact on an individual's health and safety. Although problems in this area have rarely been included on past exams, recent PE Exams have included logic problems, and it is likely that most future exams will include them.

Contents

CONTENTS

A Programmed Review For Electrical Engineering

1 Fundamental Concepts of Electrical Engineering

A review of electrical quantities, Ohm's law, circuit elements, series and parallel combination, wye–delta transformation, complex algebra, circuit element equations, transients, Laplace transform, Kirchhoff's laws, Thevenin's theorem, maximum power transfer theorem and corollary, Norton's equivalent circuit, voltage division and superposition, magnetic circuit terms, determinants, resonance, ideal transformer, and waveforms.

INTRODUCTION

This chapter summarizes the fundamental concepts of electrical engineering and serves as the foundation upon which subsequent chapters are based. The several branches of electrical engineering reviewed in the following chapters make use of these fundamental concepts in unique, applied situations. Each chapter contains additional formulas and concepts particular to the subject under discussion; however, there will always be an analog between that specialized material and the fundamentals presented in this first chapter. Therefore, the reader will want to refer to this chapter for definition of terms, conversion factors, and other information useful in solving the problems presented in subsequent chapters.

No problems are presented for solution in this chapter. It is intended as a basic review of material the reader must understand in order to solve problems in the various branches of electrical engineering contained in this text as well as on the PE Examination. Careful, thorough review of these fundamental concepts will help to assure your success in solving problems rapidly and accurately.

ELECTRICAL QUANTITIES

The basic electrical quantities are energy, power, charge, current, voltage, resistance, and electromotive force.

The basic electrical quantities are summarized below and described in more detail in the following paragraphs.

Table 1-1. Basic Electrical Quantities.

Quantity	Units	Symbol	Formula	Equivalent Units
energy (work)	joule	W	$W = \int P \, dt$	watt-second
power	watt	P	$P = \dfrac{dW}{dt}$	volt-ampere or joule/second
charge	coulomb	Q	$Q = CV$	ampere-second
current	ampere	I	$I = \dfrac{V}{R}$	coulomb/second
electrostatic pressure	volt	V	$V = IR$	joule/coulomb
resistance	ohm	R	$R = \dfrac{V}{I}$	volt/ampere

In addition to the basic units, each quantity may also be accompanied by any of the following prefixes as a scaling modifier.

Table 1-2. Scaling Modifiers.

Symbol*	Prefix	Multiplier
m	milli-	10^{-3}
μ	micro-	10^{-6}
n	nano-	10^{-9}
p	pico-	10^{-12}
K	kilo-	10^{3}
M	mega-	10^{6}
G	giga-	10^{9}

*Note that symbols for negative powers of 10 are lower case and symbols for positive powers of 10 are upper case.

Energy (Work)

Energy (W) is defined by the following formula:

$$W = \int_{t_1}^{t_2} P \, dt$$

where

W = energy (in joules) converted between times t_1 and t_2
P = power in watts
T = time in seconds

If power is steady, $W = Pt$.

Three units of energy most commonly used in electrical engineering are:

Joule or *watt-second*, defined as kinetic energy; the formula for the mechanical analog is $\frac{1}{2} mv^2$, where m is mass and v is velocity.

Kilowatt-hour, a unit used in electrical power system calculations.

Electron-volt, a unit used in calculations of electron behavior, the amount of kinetic energy acquired by an electron accelerated by a potential difference of one volt.

Table 1-3. Relations between Various Units of Energy.

watt-second = 1 joule
watt-second = 0.239 calorie
watt-second = 0.738 foot-pound
kilowatt-hour = 3413 Btu
kilowatt-hour = 1.34 horsepower-hours
kilowatt-hour = 3.6×10^6 joules
electron-volt = 1.6×10^{-9} joule
erg = 10^{-7} joule

Power

Power is the time rate of doing work. For a constant current I maintained through any load having a voltage V across it, the power is $P = IV$. With time-varying current i and voltage v, the average power is:

$$P_{av} = \frac{1}{t} \int_0^t iv \, dt$$

The units for power are obtained from energy units when the unit of time is specified. Typical units and conversion factors are:
watt = 0.239 calorie/second = 0.000948 Btu/second
joule per second = 1 watt
foot-pound per second = 1.356 watts
horsepower = 745 watts = 550 foot-pound/second

Charge

Electric charge or quantity of electricity Q is that amount of electricity passed through a circuit during a specified time interval by an electric current. The basic unit of charge is the *coulomb* and is equal to the quantity of electricity transported in one second across any cross section of a circuit by a current of one ampere. A coulomb is also equal to the charge possessed by 6.24×10^{18} electrons.

Current

Electric current is the rate of flow of electrons through a conductor. Current may be classed according to the manner in which it changes with time, including:

Direct current: a unidirectional current that may vary in amount with time but never reverses direction.

Pulsating current: a direct current that pulsates in magnitude periodically.

Continuous current: an essentially nonpulsating direct current. Unless otherwise specified, this is the type of current referred to when the term *direct current* is used.

Alternating current: a current that changes direction periodically. Normally, the net flow of current is zero. Electrons move slightly back and forth past any fixed cross section of conductor without any progression along the conductor.

The basic units of current are:

ampere
coulomb per second = 1 ampere

Voltage

Voltage refers to the quantity of energy that is gained or lost when a charge is moved from one point to another in an electric circuit. This quantity is called *potential difference* and is measured in *volts*. The potential difference between points *a* and *b* is expressed by the formula:

$$V_{ab} = \frac{W}{Q} = \frac{\text{joule}}{\text{coulomb}}$$

where

Q = coulombs transferred
W = joules lost or gained by Q during transfer

Another way of expressing potential difference is by considering *dw*, the gain or loss in energy, and *dq*, the charge of a very small particle. Thus,

$$V_{ab} = \frac{dw}{dq}$$

If *n* charged particles are transferred from *a* to *b* in an interval of time *dt*, the total charge is *ndq* and the total energy is *ndw*. Thus,

$$V = \frac{ndw}{ndq} = \frac{ndw/dt}{ndq/dt} = \frac{\text{watts}}{\text{amperes}} = \frac{P}{I}$$

4

Electromotive Force

Electromotive force (emf) relates to the physical process by which energy is changed from nonelectrical form into electrical form, in the presence of a force which tends to separate electric charges. This action occurs in electric generators, in thermocouples, and in chemical cells. Emf is defined the same way as potential difference:

$$E = \frac{dw}{dq} = \frac{P}{I} = V$$

The difference between the two is generally defined as follows:

E relates to energy being given up or generated.
V relates to energy being consumed, or a potential difference between two terminals.

Resistance

The ratio of potential difference to current is called *resistance* (R). Thus,

$$R = \frac{V}{I}$$

This is an expression of *Ohm's law*; the basic unit of resistance is the *ohm*. The resistor is a common circuit element having resistance. There are also other circuit elements, defined later, which react to impede current flow; these too are expressed quantitatively in ohms.

OHM'S LAW

Ohm's law defines the relationship between the voltage across and the current through a resistance. Thus,

$$V = IR$$
$$I = V/R$$
$$R = V/I$$

The resistance must be constant under all conditions. In ac circuits other reactive elements, such as capacitance and inductance, may also be present. Ohm's law applies to these elements as well, as long as they are constant.

Ohm's law does not hold if the circuit elements are nonlinear or the current is not steady state. Non–steady state current conditions are discussed in this chapter under "transients."

CIRCUIT ELEMENTS DEFINED

The three basic passive circuit elements are the resistor, the inductor, and the capacitor.

Characteristics of the three basic passive circuit elements are summarized in Table 1-4 and defined in the following paragraphs. Discussion of circuit elements is based on Ohm's law.

Table 1-4. Circuit Element Characteristics.

Element	Schematic Symbol	Current Through	Voltage Drop	Power Dissipation	Stored Energy	Units
resistor		$I = \dfrac{V}{R}$	$V = IR$	V^2/R or $I^2 R$	zero	ohm Ω, KΩ, MΩ
inductor		$I = \dfrac{\phi}{L}$ $= \dfrac{1}{L}\int V\,dt$	$V = -L\dfrac{dI}{dt}$	zero	$W = \frac{1}{2}LI^2$	henry h, mh, μh
capacitor		$I = C\dfrac{dV}{dt}$	$V = \dfrac{1}{C}\int I\,dt$	zero	$W = \frac{1}{2}CV^2$	farad f, μf, pf

$L = N\dfrac{d\phi}{dt}$

Resistor

A resistor is a device in which the flow of electric current always produces heat and nothing else. Since in a resistor all the electric power is consumed in generating heat, the current through a resistor must always flow from plus to minus. If the current is reversed, the potential difference must reverse; if the current is zero, the potential difference must be zero.

Inductor

An inductor is a device capable of storing and giving up magnetic energy. A pure inductor dissipates no heat or energy. When the current i in an electric circuit varies with time, a self-induced voltage is produced according to the formula:

$$V = -L\,\frac{di}{dt} \quad \{\text{polarity is such as to oppose } change \text{ in current}\}$$

in which the proportionality constant L is the *inductance* or the *coefficient of self induction*, expressed in henries. In ac circuits inductance is termed reactive or imaginary, as opposed to resistance, which is real. This subject is discussed further in this chapter under "complex algebra."

Capacitor

A capacitor is a device capable of storing or giving up electric energy. The charge on a capacitor depends upon the voltage across it and its capacitance, as described by the formula:

$$Q = CV$$

where

Q = charge in coulombs
C = capacitance in farads
V = voltage in volts

In ac circuits, capacitance, like inductance, is termed reactive or imaginary. In writing circuit equations, capacitance is given an imaginary sign 180° out of phase with any circuit inductance.

CIRCUIT ELEMENT VALUES

Values of the three basic circuit elements may be calculated from the information given below.

Resistance

The resistance of a section of conductor of uniform cross section is:

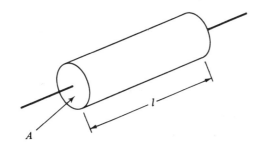

$$R = \frac{\rho l}{A}$$

where

A = cross-sectional area (square meters or circular mils*)
l = length (meters or feet)
R = resistance (ohms)

ρ = resistivity of material (ohm-meters) = $\dfrac{1}{\text{conductivity}}$

Table 1-5.   Values of Resistivity for Typical Conductive Materials.

Material	Resistivity, ohm-meters	Resistivity, $\dfrac{\text{ohm-cir mils}}{\text{ft}}$
aluminum	2.6×10^{-8}	17
copper	1.7×10^{-8}	10
cast iron	9.7×10^{-8}	58
lead	22×10^{-8}	132
silver	1.6×10^{-8}	9.9
steel	$(11–90) \times 10^{-8}$	66–540
tin	11×10^{-8}	69
nichrome	100×10^{-8}	602

*The area of a circle one mil (0.001 inch) in diameter is one circular mil; the area of any circle in circular mils equals the square of its diameter in mils.

Inductance

The inductance of a coil is:

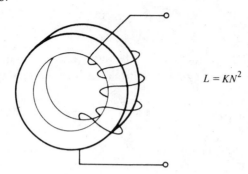

$$L = KN^2$$

where

  N = number of turns
  L = inductance in henries
  K = a constant dependent upon geometry and material

Formulas for specific shapes are as follows:

Solenoid with Nonmagnetic Core

$$L = \frac{4\pi^2 \times 10^{-7} \, N^2 r^2}{S} \text{ (henries)}$$

where

  S = length of solenoid
  r = radius of solenoid
  N = number of turns

Toroid of Rectangular Section with Nonmagnetic Core

$$L = 2N^2 b \left(\ln \frac{g+a}{g-a} \right) 10^{-9} \text{ (henries)}$$

where

  b = axial length of core, in cm
  g = mean diameter of ring, in cm
  a = radial depth of core, in cm
  N = number of turns

Toroid of Circular Section with Nonmagnetic Core

$$L = 2\pi N^2 (g - \sqrt{g^2 - d^2}) 10^{-9} \text{ (henries)}$$

where

  d = diameter of cross section, in cm
  g = mean diameter of ring, in cm
  N = number of turns

When every flux line is linked by every stream line of electric current, as at the center of long solenoids and in toroids, and the permeability of the entire circuit is independent of current, the constant K is replaced by $\mathfrak{R}$ which is the reluctance of the complete magnetic circuit.

If two coils exist in a series circuit and there is mutual coupling inductance M between them, the total inductance is either:

$L = L_1 + L_2 + 2M$ for additive mutual inductance, or
$L = L_1 + L_2 - 2M$ for bucking (subtractive) mutual inductance.

Capacitance

The capacitance of two parallel plates is:

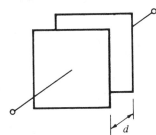

$$C = \frac{\epsilon_r \epsilon_v A}{d} = \frac{DA}{\mathcal{E}d}$$

where

$$\epsilon_v = \frac{10^{-9}}{36\pi} \text{ farads/meter} = 8.84 \text{ pf/m}$$

ϵ_r = relative permittivity of dielectric (dielectric constant)
A = common area (square meters)
d = plate spacing (meters)
C = capacitance (farads)
D = electric flux density (coulombs per square meter)
$\mathcal{E}$ = electric field intensity (volts per meter)

COULOMB'S LAW

Coulomb's law is often used as the basis for developing the theory of electrostatic fields. It defines the force on a charge in an electric field or the force between two charges.

In the field of a charge $+Q$ concentrated at a point, the total flux, ϕ, through a sphere of radians, r, is:

$$\phi = 4\pi r^2 D$$

But flux is also equal to the charge, Q, so:

$$Q = 4\pi r^2 D$$

and $D = Q/4\pi r^2$, the flux density at distance r caused by a point charge, Q. Electric field intensity, $\mathcal{E}$, at this radius is:

$$\mathcal{E} = \frac{Q}{4\pi \epsilon r^2}$$

If a charge, Q_2, is placed in the field of a charge, Q_1, the force, F, on Q_2 is:

$$F = \mathcal{E}Q_2 = \frac{Q_1 Q_2}{4\pi\epsilon r^2}$$

where

F = force in newtons
Q = charge in coulombs
ϵ = permittivity of the medium = 8.85×10^{-12} for air
r = distance between charged particles in meters
  6.28×10^{18} electrons have the charge of one coulomb
  1 electron has a charge of 1.6×10^{-19} coulomb

SERIES AND PARALLEL COMBINATIONS

The effect of connecting resistors, inductors, or capacitors in series or in parallel with their own kind is shown in Table 1-6.

In a series circuit, the same current flows through each and every element; the total voltage drop is the algebraic sum of all of the individual voltage drops.

In a parallel circuit, the same voltage appears across each and every element; the total current is the algebraic sum of all of the individual currents.

Table 1-6. Series–Parallel Combinations.

Circuit Element	Series	Parallel
Resistor R_1, R_2	$R = R_1 + R_2$	$R = \dfrac{R_1 R_2}{R_1 + R_2}$ or $R = \left(\dfrac{1}{R_1} + \dfrac{1}{R_2}\right)^{-1}$
Inductor L_1, L_2	$L = L_1 + L_2$	$L = \dfrac{L_1 L_2}{L_1 + L_2}$ or $L = \left(\dfrac{1}{L_1} + \dfrac{1}{L_2}\right)^{-1}$
Capacitor C_1, C_2	$C = \dfrac{C_1 C_2}{C_1 + C_2}$ or $C = \left(\dfrac{1}{C_1} + \dfrac{1}{C_2}\right)^{-1}$	$C = C_1 + C_2$

WYE-DELTA (Y-Δ) TRANSFORMATION

In solving circuit networks, it is often desirable to reduce them to a simpler form. The procedure is sometimes complicated, however, because the connections of circuit elements are such that the network cannot be reduced by series and parallel combinations. The Y-Δ transformation sometimes is a useful technique; it is summarized below.

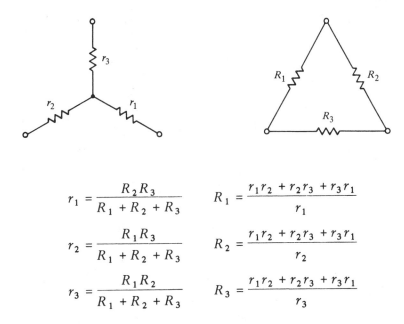

$$r_1 = \frac{R_2 R_3}{R_1 + R_2 + R_3} \qquad R_1 = \frac{r_1 r_2 + r_2 r_3 + r_3 r_1}{r_1}$$

$$r_2 = \frac{R_1 R_3}{R_1 + R_2 + R_3} \qquad R_2 = \frac{r_1 r_2 + r_2 r_3 + r_3 r_1}{r_2}$$

$$r_3 = \frac{R_1 R_2}{R_1 + R_2 + R_3} \qquad R_3 = \frac{r_1 r_2 + r_2 r_3 + r_3 r_1}{r_3}$$

COMPLEX ALGEBRA

The following is a brief review of complex algebraic notation as applied to electrical engineering.

$j = \sqrt{-1}$ = imaginary component, referred to as the j-operator

$(a + jb) + (c + jd) = (a + c) + j(b + d)$

complex conjugate of $a + jb = \overline{a + jb} = a - jb$

$(a + jb)(c + jd) = (ac - bd) + j(ad + bc)$

$\dfrac{a + jb}{c + jd} = \dfrac{(a + jb)(c - jd)}{c^2 + d^2} = \dfrac{(ac + bd) + j(bc - ad)}{c^2 + d^2}$

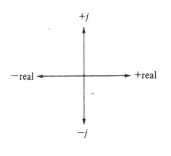

$A\underline{/\theta} = A \cos\theta + jA \sin\theta = Ae^{j\theta}$

$(A\underline{/\theta})(B\underline{/\alpha}) = AB\underline{/\theta + \alpha}$

$\dfrac{A\underline{/\theta}}{B\underline{/\alpha}} = \dfrac{A}{B}\underline{/\theta - \alpha}$

$A\underline{/\theta} + B\underline{/\alpha} = (A \cos\theta + B \cos\alpha) + j(A \sin\theta + B \sin\alpha) = C\underline{/\delta}$

where

$$C = \sqrt{(\Sigma \text{ real})^2 + (\Sigma \text{ imaginary})^2}$$

$$\delta = \text{arc tan} \left[\frac{\Sigma \text{ imaginary}}{\Sigma \text{ real}} \right]$$

$$(A\underline{/\theta})^m = A^m \underline{/m\theta}$$

$$(A\underline{/\theta})^{1/m} = A^{1/m} \left/ \frac{\theta}{m} + \frac{2\pi n}{m} \right.$$

where $n = 0, 1, 2, \ldots, (m - 1)$.

COMPLEX NOTATION

The several forms of complex notation may be summarized as follows:

    rectangular form: $Z = R \pm jX$
    polar form: $Z = |Z|\underline{/\pm\theta}$, where $\theta = \text{arc tan } X/R$ in degrees
    exponential form: $Z = |Z| e^{\pm j\theta}$, where θ is expressed in radians
    trigonometric form: $Z = |Z| (\cos \theta \pm j \sin \theta)$

CIRCUIT ELEMENT EQUATIONS

In performing circuit analysis, a convenient notational format is needed to manipulate the three circuit elements.

In a purely inductive ac circuit the current lags the voltage by 90°. In a purely capacitive circuit, the current leads the voltage by 90°. The sum of a resistance and an inductance is $R + jX_L$ and is written as an impedance:

$$Z = R + jX_L \text{ ohms}$$

The sum of a resistance and a capacitance is $R - jX_C$ and is written as an impedance:

$$Z = R - jX_C \text{ ohms}$$

These two equations can be manipulated as required using complex algebraic relationships. The actual values of X_L and X_C may be calculated from the following formulas:

$$X_L = 2\pi f L = \omega L$$

$$X_C = \frac{1}{2\pi f C} = \frac{1}{\omega C}$$

In complex notation, these are written:

$$jX_L = j\omega L$$

$$jX_C = j\frac{1}{\omega C} = -\frac{1}{j\omega C}$$

In certain cases it is more convenient to work with the reciprocal of impedance and its real and imaginary components. Such cases include parallel circuits and current (rather than voltage) sources. The following relationships hold:

$$\frac{1}{Z} = Y = \text{admittance} = \frac{1}{\text{impedance}}$$

$$\frac{1}{R} = G = \text{conductance} = \frac{1}{\text{resistance}}$$

$$-\frac{1}{X} = B = \text{susceptance} = \frac{-1}{\text{reactance}}$$

Thus,

$$Y = G + jB \text{ mhos}$$

$$Z = R + jX \text{ ohms}$$

TRANSIENTS

In a preceding section, only circuits where Ohm's law applies were considered. These circuits are completely independent of time. In every practical circuit, however, there are two reactive elements present, no matter how small, which manifest themselves when current is changing. The effect that these two elements, inductance and capacitance, have in a transient situation is discussed below.

Single Energy Transients

Circuits containing only one type of reactive element are readily analyzed using a generalized transient response formula of the form:

$$f(t) = f_{ss} + (f_0 - f_{ss}) e^{-t/\tau}$$

where

f_{ss} = steady state final value of $f(t)$ $(t \to \infty)$
f_0 = initial value of $f(t)$ $(t = 0^+)$
τ = circuit time constant

This formula may be applied to any voltage or current in the circuit, even with initial nonzero values of inductor current or capacitor voltage, as long as only a single time constant is present (i.e., the circuit is first order).

Formulas specific to series RL and RC circuits are summarized below:

a. RL Circuit
1. Inductance is associated with the magnetic field of the current.
2. Inductance has the mathematical property $e = L \, di/dt$.
3. $i = I_{ss} + (I_0 - I_{ss}) e^{-t/\tau}$
   where
       $I_{ss} = V/R$
        $\tau = L/R$ seconds
4. Stored energy is $W = \frac{1}{2} Li^2$.

b. RC Circuit
1. Capacitance is associated with the electric field.
2. Capacitance has the mathematical property $i = C \, dv/dt$.

3. $v = V_{ss} + (V - V_{ss})\,e^{-t/\tau}$
   where $\tau = RC$ seconds
4. $i = dq/dt$
5. Charge is $Q = CV$
6. Stored energy is $W = \frac{1}{2}\,CV^2$.

Double Energy Transients

Double energy, or second order, systems are those in which energy can be stored in two separate forms. These circuits contain R, L, and C. Their analysis can become quite complex because of damped oscillations and complex equations. They are usually most easily solved using Laplace transforms.

For an RLC series circuit the loop equation for voltage is:

$$E = L\,\frac{di}{dt} + Ri + \frac{1}{C}\int_0^t i\,dt$$

Circuit Examples

Transient analysis of circuits containing reactive elements may be performed using any of the following methods:

   a. Transient response formula for single energy circuits.
   b. Differential equations.
   c. Laplace transforms.
   d. Computer solutions.

Examples of some specific circuits follow.

Example 1.   VL

$$V_L = L\,\frac{dI}{dt} = V_1, \qquad \frac{dI}{dt} = \frac{V_1}{L}$$

Example 2.   VC

$V_C = V_1$ after switch closed

$I(t) \to \infty$ upon closing switch

$I(t) \to 0$ for $t > 0^+$

Example 3. VRL

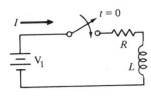

$$I(t) = I_{ss} + (I_0 - I_{ss}) e^{-t/\tau}$$

$$= \frac{V_1}{R} (1 - e^{-(R/L)t})$$

$$\left.\begin{array}{l} I_0 = 0 \\[2mm] I_{ss} = V_1/R \\[2mm] \tau = L/R \end{array}\right.$$

$$V_L = L\frac{di}{dt} = L\frac{V_1}{R}\frac{R}{L}e^{-(R/L)t} = V_1 e^{-(R/L)t}$$

Example 4. VRC

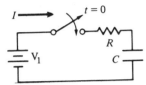

$$I(t) = I_{ss} + (I_0 - I_{ss}) e^{-t/\tau}$$

$$= \frac{V_1}{R} e^{-t/RC}$$

$$\left.\begin{array}{l} I_0 = V_1/R \\[2mm] I_{ss} = 0 \\[2mm] \tau = RC \end{array}\right.$$

$$V_C(t) = \frac{1}{C}\int_0^t I\,dt = \frac{V_1}{RC}\int_0^t e^{-t/RC}\,dt = -V_1 e^{-t/RC}\Big|_0^t = V_1(1 - e^{-t/RC})$$

Note: $V_C(t)$ may be found directly by using the transient formula

$$V_C(t) = V_{ss} + (V_0{'} - V_{ss}) e^{-t/\tau}$$

$$= V_1(1 - e^{-t/RC})$$

Example 5. VC/CR

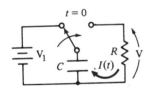

$$V_C(t) = V_{ss} + (V_0 - V_{ss}) e^{-t/\tau}$$

$$= V_1 e^{-t/RC}$$

$$\left.\begin{array}{l} V_0 = V_1 \\[2mm] V_{ss} = 0 \\[2mm] \tau = RC \end{array}\right.$$

$$I(t) = I_{ss} + (I_0 - I_{ss}) e^{-t/\tau}$$

$$= \frac{V_1}{R} e^{-t/RC}$$

$$\left.\begin{array}{l} I_0 = V_1/R \\[2mm] I_{ss} = 0 \\[2mm] \tau = RC \end{array}\right.$$

Example 6. RCC

$$W = \tfrac{1}{2} CV^2\,; \text{ for } t < 0; V_1 = V_0, V_2 = 0$$

$$I(t) = I_{ss} + (I_0 - I_{ss}) e^{-t/\tau}$$

$$= \frac{V_1}{R} e^{-t/\tau}$$

$$\left.\begin{array}{l} I_0 = V_1/R \\[2mm] I_{ss} = 0 \\[2mm] \tau = R\dfrac{C_1 C_2}{C_1 + C_2} \end{array}\right.$$

$$V_2(t) = \frac{1}{C_2}\int_0^t I\,dt = \frac{V_1}{RC_2}\int_0^t e^{-t/\tau}\,dt = \frac{-\tau V_1}{RC_2} e^{-t/\tau}\Big|_0^t = \frac{C_1 V_1}{C_1 + C_2}(1 - e^{-t/\tau})$$

Initial energy $= \frac{1}{2} C_1 V_1^2$

final energy $= \frac{1}{2}(C_1 + C_2)V_{ss}^2 = \frac{1}{2}(C_1 + C_2)\left(\frac{C_1 V_1}{C_1 + C_2}\right)^2 = \frac{1}{2}\frac{C_1 V_1^2}{\left(1 + \frac{C_2}{C_1}\right)}$

lost energy (dissipated in R) $= \frac{1}{2}C_1 V_1^2 - \frac{1}{2}\frac{C_1 V_1^2}{\left(1 + \frac{C_2}{C_1}\right)} = \frac{1}{2}\frac{C_1 V_1^2}{\left(1 + \frac{C_1}{C_2}\right)}$

If $C_1 = C_2$, then one-half of the energy is lost.

$$\frac{\text{final energy}}{\text{initial energy}} = \frac{C_1}{C_1 + C_2}$$

$$\frac{\text{lost energy}}{\text{initial energy}} = \frac{C_2}{C_1 + C_2}$$

LAPLACE TRANSFORM

The Laplace transform is a transformation technique relating time functions to frequency dependent functions of a complex variable. The Laplace transform is defined as follows:

$$\mathcal{L}[f(t)] = F(s) = \int_0^\infty f(t)\, e^{-st}\, dt$$

The inverse Laplace transform is written in the following notation:

$$\mathcal{L}^{-1}[F(s)] = f(t)$$

Table 1-7. Laplace Transforms.

Description	Laplace Transform $F(s) = \mathcal{L}f(t)$	Time Function $f(t)$
unit pulse	1	$\delta(t)$
unit step	$1/s$	$U(t)$
unit ramp	$1/s^2$	t
polynomial	$n!/s^{n+1}$	t^n
exponential	$1/(s - a)$	e^{at}
sine wave	$1/(s^2 + \omega^2)$	$\frac{1}{\omega}\sin \omega t$
cosine wave	$s/(s^2 + \omega^2)$	$\cos \omega t$
damped sine wave	$1/[(s - a)^2 + \omega^2]$	$\frac{1}{\omega}e^{at}\sin \omega t$
damped cosine wave	$(s - a)/[(s - a)^2 + \omega^2]$	$e^{at}\cos \omega t$

Example:

$$\mathcal{L}[e^{-t}] = \int_0^\infty e^{-t}e^{-st}\,dt = \frac{1e^{-(s+1)t}}{-(s+1)}\Big|_0^\infty = \frac{1}{s+1}$$

Table 1-7 lists several of the more common Laplace transforms.
Table 1-8 shows how the three basic circuit elements are symbolized in the s-domain and time domain.

Table 1-8. Circuit Element Symbology.

	$Z(s)$	$Z(j\omega)$	$X(j\omega)$
resistor	R	R	R
inductor	sL	$j\omega L$	ωL
capacitor	$\dfrac{1}{sC}$	$\dfrac{1}{j\omega C}$	$\dfrac{1}{\omega C}$

KIRCHHOFF'S LAWS

Two fundamental simple laws of the electric circuit have received the name of Kirchhoff's laws. These are stated as follows:

First Law. The amount of direct current flowing away from a point in a circuit is equal to the amount flowing to that point. (Incidentally, this law applies equally well to ac, dc, and ac–dc hybrid circuits.) In short,

$$\sum \text{currents entering a node} = 0$$

Second Law. The difference of electric potential between any two points is the same regardless of the path along which it is measured. In short,

$$\sum \text{voltage drops (or rises) about a closed path} = 0$$

The number of independent equations which can be written using Kirchhoff's first law is *one* less than the number of *nodes* or junction points at which two currents join to form a third. The number of independent equations which can be written using Kirchhoff's second law is equal to the number of branches minus the number of independent node equations; a *branch* is any section of a circuit which directly joins two nodes without passing through a third node.

THEVENIN'S THEOREM

Thevenin's Theorem is a handy tool in the solution of both dc and ac circuits. It applies only to the terminal voltage and current conditions of a linear two-terminal network not magnetically coupled to an external network. For dc circuits, it states that one can replace a two-terminal network by a voltage source E_T and a resistance R_T connected in series.

17

For example, the circuit below on the left may be converted to that shown on the right.

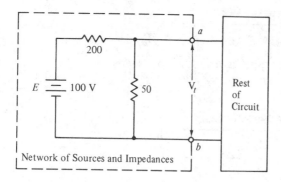 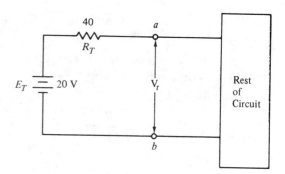

The two circuits are equivalent if:

1. E_T is made equal to V_t on open circuit; thus, both circuits have equal open circuit voltages.
2. R_T is given a value such that when the terminals $a–b$ of each are short-circuited, the current through the short circuits of both are equal.

Thus, for the above example,

$$E_T = 100 \, \frac{50}{200 + 50} = 20 \text{ volts}$$

$$R_T = \frac{200 \times 50}{200 + 50} = 40 \text{ ohms}$$

MAXIMUM POWER TRANSFER THEOREM

Using the Thevenin equivalent of a voltage source circuit, it is easy to study the characteristics of the rest of the circuit. With a "Thevenized" circuit:

$$V_t = E_T - I_t R_T$$

$$P_t = V_t I_t = E_T I_t - I_t^2 R_T$$

Maximum power occurs at a value of $I_t = \frac{1}{2} E_T / R_T$ and at a value of $V_t = E_T / 2$, so that maximum power is, $P_{\max} = (E_T)^2 / 4R_T$. This is obtained by differentiating the above equation for P_t with the respect to I_t and setting the result equal to zero. The resistance of the load, for maximum power transfer, is:

$$R_{L \max p} = \frac{V_t}{I_t} = \left(\frac{E_T}{2} \right) \left(\frac{2R_T}{E_T} \right) = R_T$$

This result demonstrates the *maximum power transfer theorem* which states that the maximum power is delivered to a load by a two-terminal linear network when that load is so adjusted that the terminal voltage is half its open-circuit value.

MAXIMUM POWER TRANSFER THEOREM COROLLARY

The maximum power transfer theorem, restated, says that maximum power will be delivered by a network to an impedance Z_R if Z_R is the complex conjugate of the Z's of the network, measured looking back into the terminals of the network.

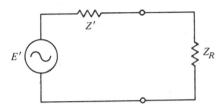

A corollary states that if only the absolute magnitude and not the angle of Z_R may be varied, then the greatest power output will be delivered from the network if the absolute magnitude of Z_R is made equal to the absolute magnitude of Z':

$$|Z_R| = |Z'|$$

The amount of power delivered by matching magnitudes will be somewhat less than the amount possible if both magnitude and angle are adjusted to the conjugate condition.

NORTON'S EQUIVALENT CIRCUIT

In the solution of many electric circuits, it is often more advantageous to consider constant-current sources rather than the constant-voltage sources thus far considered. The comparison between a Norton equivalent circuit and a Thevenin equivalent circuit for a two-port network is illustrated below.

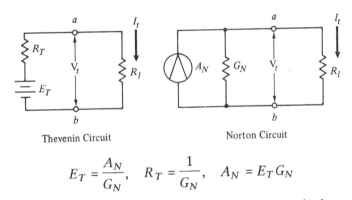

Thevenin Circuit   Norton Circuit

$$E_T = \frac{A_N}{G_N}, \quad R_T = \frac{1}{G_N}, \quad A_N = E_T G_N$$

where A_N is a constant-current source through short-circuited terminals.

$$I_t = A_N - V_t G_N$$

$$V_t = \frac{A_N}{G_N} - \frac{I_t}{G_N}$$

Note that these two sources are equivalent only in the current, voltage, and power they deliver to the terminals a–b. They are not equivalent in the amount of internal power they consume within themselves.

VOLTAGE DIVISION AND SUPERPOSITION

Two time-saving methods of analyzing circuits are illustrated below.

Voltage Division

The voltage V_l across the load terminals of a circuit may be determined in this manner:

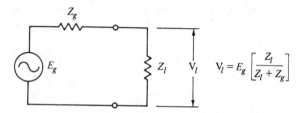

$$V_l = E_g \left[\frac{Z_l}{Z_l + Z_g} \right]$$

Superposition

The response of a linear network to a number of excitations applied simultaneously is equal to the sum of the responses of the network when each excitation is applied individually.

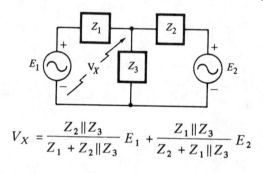

$$V_X = \frac{Z_2 \| Z_3}{Z_1 + Z_2 \| Z_3} E_1 + \frac{Z_1 \| Z_3}{Z_2 + Z_1 \| Z_3} E_2$$

where

$$Z_2 \| Z_3 = \frac{Z_2 Z_3}{Z_2 + Z_3}$$

MAGNETIC CIRCUIT TERMS

A typical magnetic circuit diagram appears like this:

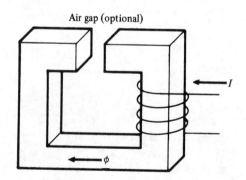

Air gap (optional)

In RMKS units, the following terms apply:

F = magnetomotive force (amp turns)

ϕ = flux (webers)

H = magnetic field intensity (amp turns/meter)

B = flux (magnetic field) density (weber/square meter)

μ_r = permeability (dimensionless)

$\mathcal{R}$ = reluctance

I = current (amps)

N = no. of complete turns about core

A = cross-sectional area (square meters)

l = length of magnetic path (meters)

μ_v = permeability of free space = $4\pi \times 10^{-7}$ webers/amp turn–meter

Some relationships between terms include:

$\phi = F/\mathcal{R}$ (similar to $I = V/R$)

$F = NI$

$H = NI/l$

$V = N\,d\phi/dt$

$\mathcal{R} = l/\mu\,\mu_v A$

$\mu = B/H$

$\mathcal{R}_{tot} = \mathcal{R}_1 + \mathcal{R}_2$ (reluctances combine like resistances)

Table 1-9 expresses these terms in different units and illustrates their interrelationships.

Table 1-9. Magnetic Circuit Symbols, Units, and Interrelationships.

Symbol	RMKS Units	CGS Units	CGS Equation	CGS Units / RMKS Units	RMKS Equation
F	amp turn	gilbert	$F = 0.4NI$	1.257	$F = NI$
H	amp turn/meter	oersted	$H = F/l$	0.01257	$H = F/l$
ϕ	weber	maxwell or line	$\phi = F/\mathcal{R}$	10^8	$\phi = F/\mathcal{R}$
B	weber/meter^2 or tesla	gauss	$B = \phi/A$	10^4	$B = \phi/A$
$\mathcal{R}$			$\mathcal{R} = \dfrac{l}{A\mu_r} = \dfrac{Hl}{BA}$	1.257×10^{-8}	$\mathcal{R} = \dfrac{l}{A\mu_r\mu_v}$
A	meter^2	cm^2		10^4	
l	meter	cm		10^2	

DETERMINANTS

In solving circuit networks, node and loop equations are often written using Kirchhoff's laws. This results in n equations with n unknowns, where n is usually 2 or 3. An easy way of solving for the unknowns is to use determinants when the equations for voltage, current, and impedance are properly ordered. The following are the determinant solutions to second-order ($n = 2$) and third-order ($n = 3$) sets:

Second Order

Circuit equations:

$$V_1 = Z_A I_1 + Z_B I_2$$
$$V_2 = Z_C I_1 + Z_D I_2$$

Determinant form:

$$\begin{bmatrix} V_1 \\ V_2 \end{bmatrix} = \begin{bmatrix} Z_A Z_B \\ Z_C Z_D \end{bmatrix} \begin{bmatrix} I_1 \\ I_2 \end{bmatrix}$$

Determinant solution:

$$I_1 = \frac{\begin{bmatrix} V_1 \\ V_2 \end{bmatrix} \begin{bmatrix} Z_B \\ Z_D \end{bmatrix}}{\begin{bmatrix} Z_A \\ Z_C \end{bmatrix} \begin{bmatrix} Z_B \\ Z_D \end{bmatrix}} = \frac{V_1 Z_D - V_2 Z_B}{Z_A Z_D - Z_B Z_C} = \frac{V_1 Z_D - V_2 Z_B}{D}$$

$$I_2 = \frac{\begin{bmatrix} Z_A \\ Z_C \end{bmatrix} \begin{bmatrix} V_1 \\ V_2 \end{bmatrix}}{D} = \frac{V_2 Z_A - V_1 Z_C}{D}$$

Third Order

Circuit equations:

$$V_1 = Z_A I_1 + Z_B I_2 + Z_C I_3$$
$$V_2 = Z_D I_1 + Z_E I_2 + Z_F I_3$$
$$V_3 = Z_G I_1 + Z_H I_2 + Z_I I_3$$

Determinant form:

$$\begin{bmatrix} V_1 \\ V_2 \\ V_3 \end{bmatrix} = \begin{bmatrix} Z_A & Z_B & Z_C \\ Z_D & Z_E & Z_F \\ Z_G & Z_H & Z_I \end{bmatrix} \begin{bmatrix} I_1 \\ I_2 \\ I_3 \end{bmatrix}$$

Determinant solution:

$$I_1 = \frac{\begin{bmatrix} V_1 & Z_B & Z_C \\ V_2 & Z_E & Z_F \\ V_3 & Z_H & Z_I \end{bmatrix}}{D}$$

where

$$D = \begin{bmatrix} Z_A & Z_B & Z_C \\ Z_D & Z_E & Z_F \\ Z_G & Z_H & Z_I \end{bmatrix}$$

$$= Z_A Z_E Z_I + Z_B Z_F Z_G + Z_C Z_D Z_H - Z_G Z_E Z_C - Z_H Z_F Z_A - Z_I Z_D Z_B$$

Therefore:

$$I_1 = \frac{V_1 Z_E Z_I + V_3 Z_B Z_F + V_2 Z_H Z_C - V_1 Z_H Z_F - V_2 Z_B Z_I - V_3 Z_E Z_C}{D}$$

Similarly:

$$I_2 = \frac{\begin{bmatrix} Z_A & V_1 & Z_C \\ Z_D & V_2 & Z_F \\ Z_G & V_3 & Z_I \end{bmatrix}}{D}$$

$$= \frac{V_1 Z_F Z_G + V_2 Z_A Z_I + V_3 Z_D Z_C - V_1 Z_D Z_I - V_2 Z_C Z_G - V_3 Z_F Z_A}{D}$$

$$I_3 = \frac{\begin{bmatrix} Z_A & Z_B & V_1 \\ Z_D & Z_E & V_2 \\ Z_G & Z_H & V_3 \end{bmatrix}}{D}$$

$$= \frac{V_1 Z_D Z_H + V_2 Z_B Z_G + V_3 Z_A Z_E - V_1 Z_G Z_E - V_2 Z_H Z_A - V_3 Z_B Z_D}{D}$$

RESONANCE

Resonance is defined as the property of cancellation of reactance when inductive and capacitive reactances are in series, or cancellation of susceptance when in parallel. Under resonant conditions, reactive circuits operate at *unity power factor* with current and voltage in phase ($\theta = 0^0$ and $\cos \theta = 1$). Basic electrical engineering texts deal at length with resonance. Some of the basic formulas are delineated here.

Figure of Merit, Q

$$Q = 2\pi \times \frac{\text{maximum energy stored per cycle}}{\text{energy dissipated per cycle}}$$

For an inductor:

$$Q = \frac{\omega L}{R_s}$$

$$Q = \frac{R_p}{\omega L}$$

23

For a capacitor:

$$Q = \omega C R_p$$

$$Q = \frac{1}{\omega C R_s}$$

For series RLC:

$$Q = \frac{\text{series reactance}}{\text{series resistance}}$$

$$Q = \frac{\omega_r L}{R} = \frac{1}{\omega_r RC}$$

where ω_r is at the resonant frequency.

For parallel-to-series conversion, the following derivation is applied at a specific frequency:

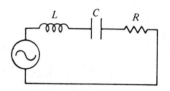

$$Z = \frac{R(jX)}{R + jX} = \frac{jRX(R - jX)}{R^2 + X^2} = \frac{RX^2 + jXR^2}{R^2 + X^2}$$

Therefore,

$$Q = \left[\frac{XR^2}{R^2 + X^2}\right]\left[\frac{R^2 + X^2}{RX^2}\right] = \frac{R}{X}$$

Series Resonance

At resonant frequency,

$$\omega_r L = \frac{1}{\omega_r C}, \quad \omega_r^2 LC = 1, \quad \omega_r = \frac{1}{\sqrt{LC}}$$

$$f_r = \frac{\omega_r}{2\pi} = \frac{1}{2\pi\sqrt{LC}}$$

$$Q = \frac{\omega_r L}{R} = \frac{1}{R}\sqrt{\frac{L}{C}}$$

The *bandwidth B* of a resonant circuit is defined as the width of the resonance curve between the two frequencies at which power in the circuit is one-half maximum power.

For a complete circuit including R, L, C, generator, and load,

$$B = \Delta f = \frac{f_r}{Q}$$

24

For matched conditions in which generator resistance equals the remainder of the circuit resistance,

$$B = \frac{2}{Q} f_r$$

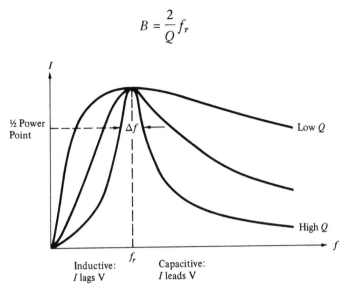

At half power point, $R = X = |X_L - X_C|$.
Series resonance yields high current and low impedance (pure resistance).

Parallel Resonance (Antiresonance)

For antiresonance, the circuit must have unity power factor. Generator current is a minimum, and impedance is a maximum. The capacitor is assumed to have no associated shunt resistance.

$$R_{ar} = \frac{L}{CR}$$

$$f_{ar} = \frac{1}{2\pi} \sqrt{\frac{1}{LC} - \frac{R^2}{L^2}}^* \approx \frac{1}{2\pi} \sqrt{\frac{1}{LC}}$$

For matched conditions, where $R_g = R_{ar}$:

$$B = \frac{2}{Q} f_{ar} \quad \text{(same as for series circuit)}$$

*This is the same expression as for a series resonant circuit except for the small second term under the radical. Resonance is not possible for values of R that make $R^2/L^2 > 1/LC$. This contrasts with the series circuit, which can be resonant for all values of R.

Rearranging the formula for f_{ar},

$$f_{ar} = \frac{1}{2\pi}\sqrt{\frac{1}{LC}}\sqrt{1 - \frac{R^2C}{L}} = \frac{1}{2\pi}\sqrt{\frac{1}{LC}}\sqrt{1 - \frac{1}{Q^2}}$$

where

$$Q = \frac{1}{R}\sqrt{\frac{L}{C}}$$

for the circuit to the right of terminals a, b. If $Q > 10$, the error by neglecting the radical $\sqrt{1 - (1/Q^2)}$ is less than 1%. This radical shows that resonance is not possible if $Q < 1$.

Impedance Transformation

Using the principles of resonance, two reactances of opposite sign may be arranged as an L-section to transform, at a single frequency, a load resistance R to provide a matched load R_{in} for the generator, as shown in the two cases below.

For $R < R_g$,

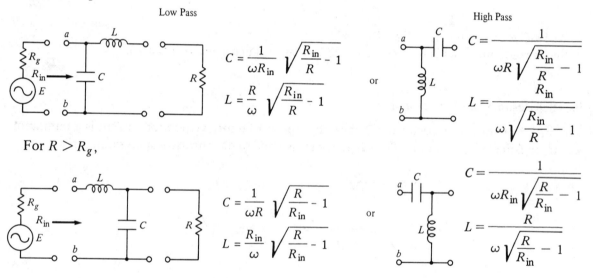

Low Pass

$$C = \frac{1}{\omega R_{in}}\sqrt{\frac{R_{in}}{R} - 1}$$

$$L = \frac{R}{\omega}\sqrt{\frac{R_{in}}{R} - 1}$$

or

High Pass

$$C = \frac{1}{\omega R\sqrt{\dfrac{R_{in}}{R} - 1}}$$

$$L = \frac{R_{in}}{\omega\sqrt{\dfrac{R_{in}}{R} - 1}}$$

For $R > R_g$,

$$C = \frac{1}{\omega R}\sqrt{\frac{R}{R_{in}} - 1}$$

$$L = \frac{R_{in}}{\omega}\sqrt{\frac{R}{R_{in}} - 1}$$

or

$$C = \frac{1}{\omega R_{in}\sqrt{\dfrac{R}{R_{in}} - 1}}$$

$$L = \frac{R}{\omega\sqrt{\dfrac{R}{R_{in}} - 1}}$$

IDEAL TRANSFORMER

An ideal transformer is one in which there are no internal losses. In analyzing circuits, it is sometimes justifiable to ignore these losses. The following are some relationships for an ideal two-winding transformer:

turns ratio, $a = \dfrac{N_p}{N_s} = \dfrac{V_p}{V_s}$

$$V_pI_p = V_sI_s$$

$$\frac{I_s}{I_p} = \frac{N_p}{N_s}$$

$$I_s = \frac{V_s}{Z_l} = \frac{V_g}{Z_l} \frac{N_s}{N_p}$$

$$I_p = \frac{N_s}{N_p} I_s = \frac{V_g}{Z_l} \left(\frac{N_s}{N_p}\right)^2 = \frac{V_g}{Z_l \left(\frac{N_p}{N_s}\right)^2}$$

Secondary load reflects into primary multiplied by $(N_p/N_s)^2$.

WAVEFORMS

It is often desired to determine the RMS value or average value of a periodic waveform. The following formulas are used:

RMS Value

a. General Periodic Waveform
The root mean square (also called virtual or effective) value of a *periodic* waveform is given by:

$$V_{RMS} = \sqrt{\frac{1}{T} \int_0^T v^2(t)\, dt}$$

and is a direct measure of heating value upon a resistive load.

Example

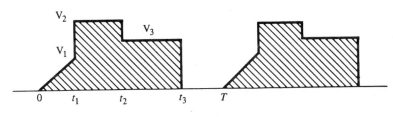

$$V_{RMS}^2 = \frac{1}{T}\left[\int_0^{t_1}\left(V_1 \frac{t}{t_1}\right)^2 dt + \int_{t_1}^{t_2} V_2^2\, dt + \int_{t_2}^{t_3} V_3^2\, dt + \int_{t_3}^T 0\, dt\right]$$

b. Sine Wave
For sine wave of $v(t) = V_{max} \sin \omega t$,

$$V_{RMS} = \frac{V_{max}}{\sqrt{2}} = 0.707 V_{max}, \quad \text{for one complete cycle}$$

c. RMS Value of Composite Waveform
$$V_{RMS} = \sqrt{V_{1\,RMS}^2 + V_{2\,RMS}^2 + V_{3\,RMS}^2 + \cdots + V_{n\,RMS}^2}$$

where every term is at a different frequency (including dc).

27

Average Value

The average value of any ac wave which is symmetrical about the zero axis is zero. However, when average value is applied to alternating quantities, it usually means the average of either the positive or negative loop of the wave. Since the average ordinate multiplied by the base is equal to the area under the curve,

$$\text{average value} = \frac{2}{T} \int_0^{T/2} i \, dt$$

where T is one cycle. This equation is applicable only when the wave passes through zero at the time $t = 0$. For any other condition, the time t_1 at which the instantaneous value of the wave is zero must be determined and the average value found from:

$$\frac{2}{T} \int_{t_1}^{t_1 + (T/2)} i \, dt$$

The average value of a sinusoid over one-half cycle is:

$$I_{av} = \frac{2}{\pi} I_{max} = 0.637 I_{max}$$

More generally, the average value of one period of any periodic waveform is:

$$\text{Average value} = \frac{\text{net positive area}}{\text{base}} = \text{dc value of } f(t)$$

$$= \frac{1}{T} \int_0^t f(t) \, dt,$$

where T is one period or cycle.

Example 1.   Unsymmetrical Square Wave

$$V_{dc} = \frac{1}{t_3} \left[\int_0^{t_1} V_1 \, dt + \int_{t_1}^{t_2} -V_2 \, dt + \int_{t_2}^{t_3} 0 \, dt \right]$$

$$= \frac{V_1 t_1 - V_2 (t_2 - t_1)}{t_3}$$

Example 2.   Symmetrical Sine Wave

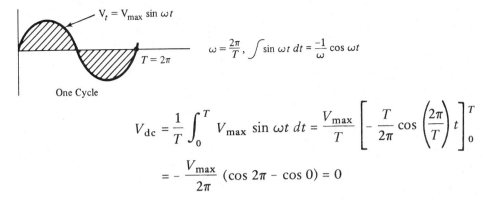

$$V_{dc} = \frac{1}{T} \int_0^T V_{max} \sin \omega t \, dt = \frac{V_{max}}{T} \left[-\frac{T}{2\pi} \cos\left(\frac{2\pi}{T}\right) t \right]_0^T$$

$$= -\frac{V_{max}}{2\pi} (\cos 2\pi - \cos 0) = 0$$

Example 3.   Rectified Half Wave

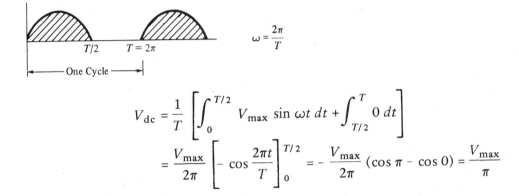

$$V_{dc} = \frac{1}{T} \left[\int_0^{T/2} V_{max} \sin \omega t \, dt + \int_{T/2}^T 0 \, dt \right]$$

$$= \frac{V_{max}}{2\pi} \left[-\cos \frac{2\pi t}{T} \right]_0^{T/2} = -\frac{V_{max}}{2\pi} (\cos \pi - \cos 0) = \frac{V_{max}}{\pi}$$

Example 4.   Rectified Full Wave

$$V_{dc} = \frac{1}{T} \int_0^T V_{max} \sin \omega t \, dt = -\frac{V_{max}}{\pi} \left[\cos \frac{\pi}{T} t \right] = \frac{2V_{max}}{\pi}$$

2 Basic Circuits

Problems related to capacitor energy and charge; two capacitor charge transfer; power, energy, and charge; series-parallel resistance; series resonance; impedance transformation; wye-delta transformation; circuit network-loop current analysis; Thevenin circuit-maximum power; ideal transformer-maximum power; ammeters; RL transient; double energy transient; series-parallel resonant filter; and waveform.

INTRODUCTION

This chapter contains basic circuit problems with solutions which use one or more of the fundamental concepts presented in Chapter 1.
   FOLLOW THESE STEPS THROUGHOUT THE REMAINDER OF THE TEXT:

1. Place a sheet of paper over the solution section of each problem.
2. Read the problem carefully.
3. Work the problem on scratch paper. You may refer to concepts and formulas contained in Chapter 1 if necessary. (Use the index to locate needed information rapidly.)
4. Evaluate your solution to the problem by comparing it to that given in the text.
5. If your solution is correct, go to the next problem; if it is incorrect, review the material indicated at the end of the solution.
6. Rework the problem.

RESISTANCE

The problem in this section illustrates the basics of combining series and parallel resistances into a single equivalent resistance.

PROBLEM 2-1. SERIES-PARALLEL RESISTANCE

Find the equivalent resistance of the resistance network shown below.

COVER
THE
SOLUTION

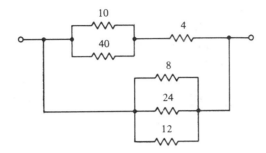

WORK THE
PROBLEM

EVALUATE   Evaluate your answer by comparing it to the solution below.

Solution:

By combining parallel resistors, the network reduces to:

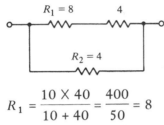

$$R_1 = \frac{10 \times 40}{10 + 40} = \frac{400}{50} = 8$$

$$R_2 = \left[\frac{1}{8} + \frac{1}{24} + \frac{1}{12}\right]^{-1} = 4$$

This circuit finally reduces to:

$$R = \frac{(8 + 4)\,(4)}{8 + 4 + 4} = \frac{48}{16} = 3 \text{ ohms} \qquad \text{(answer)}$$

If your answer is correct, go on to the next section.
If your answer is not correct, review p. 10.

WORK, ENERGY, AND POWER

The problem in this section illustrates the application of three fundamental engineering parameters.

PROBLEM 2-2. WORK, ENERGY, AND POWER

A hoist driven by a three-phase electric motor has a 4000-pound load capacity.
  a. How much energy in joules and power in KW is required to raise a full capacity load 100 feet in one minute?
  b. What motor capacity in HP is required to drive the hoist if the hoisting machinery is only 75% efficient?

Solution:

a. $W = \int_0^t P\,dt = Pt = \left[\dfrac{\text{lb-ft}}{\text{sec}}\right]\text{sec} = \text{lb-ft}$

1 watt-second = 1 joule = 0.738 lb-ft

(answer) $W = 4000\text{ lb} \times 100\text{ ft} \times \dfrac{\text{joule}}{0.738\text{ lb-ft}} = 5.42 \times 10^5\text{ joule}$

$P = \dfrac{W}{t} = \dfrac{5.42 \times 10^5\text{ joule}}{1\text{ min} \times 60\text{ sec/min}} \times \dfrac{\text{watt-sec}}{\text{joule}}$

(answer) $= 9033\text{ W or } 9.033\text{ KW}$

(answer) b. $P = 9.033\text{ KW} \times \dfrac{\text{HP}}{0.745\text{ KW}} \times \dfrac{1}{0.75} = 16.17\text{ HP}$

If your answers are correct, go on to Problem 2-3.
If your answers are not correct, review p. 3.

COULOMB'S LAW

The problem in this section illustrates the application of Coulomb's Law, as defined in Chapter 1.

PROBLEM 2-3. COULOMB'S LAW

Find the force between two electrons in air separated by a distance of 200 cm.

Solution:

$$F = \frac{Q_1 Q_2}{4\pi\epsilon r^2} = \frac{(1.6 \times 10^{-19})^2}{4\pi \times 8.85 \times 10^{-12} \times 2^2}$$

(answer) $= 5.75 \times 10^{-29}\text{ newton}$

If your answer is correct, go on to Problem 2-4.
If your answer is not correct, review pp. 8–9.

NETWORKS

The problems in this section illustrate a few aspects of network analysis and amplification. The study of lattice and other types of networks as illustrated in Chapter 8 of Reference Data for Radio Engineers[36] is recommended.

PROBLEM 2-4. WYE-DELTA TRANSFORMATION

The circuit below is a Wheatstone bridge with a resistance in place of the galvanometer. Find the current supplied by the battery.

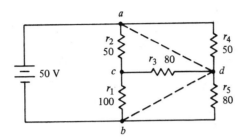

Solution:

Using the wye–delta transformation, the network may be reduced as follows:

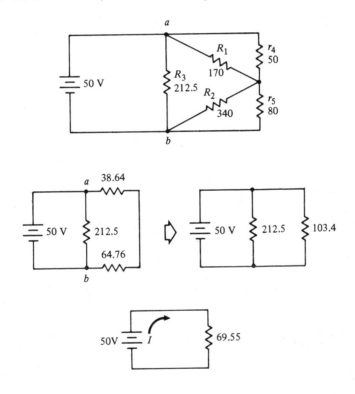

Solving for current:

$$I = \frac{V}{R} = \frac{50}{69.55} = 0.72 \text{ amp}$$

(answer)

If your answer is correct, go on to Problem 2-5.
If your answer is not correct, review p. 11.

PROBLEM 2-5. SCHERING BRIDGE

The value of an unknown capacitor is to be measured using a Schering bridge. If the values of the bridge capacitors and resistors are as shown below, what is the value of the unknown capacitor, C_x, and its series resistance, R_x?

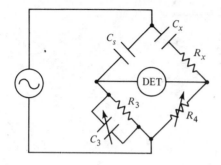

$C_s = 1000$ pf
$C_3 = 10$ pf
$R_3 = 1$ MΩ
$R_4 = 1$ KΩ

Solution:

For balanced conditions,

$$Z_1 = -j\frac{1}{\omega C_s}, \quad Z_4 = R_4, \quad Y_3 = \frac{1}{R_3} + j\omega C_3$$

$$Z_x = Z_1 Z_4 Y_3$$

$$Z_x = \left[R_x - j\frac{1}{\omega C_x}\right] = \left[-j\frac{1}{\omega C_x}\right]\left[R_4\right]\left[\frac{1}{R_3} + j\omega C_3\right] = \frac{C_3 R_4}{C_s} - j\frac{R_4}{\omega C_s R_3} = R_x - jX_{C_x}$$

Evaluating reals and imaginaries,

(answer)
$$C_x = C_s\frac{R_3}{R_4} = [10^{-9}]\frac{10^6}{10^3} = 10^{-6} \text{ or } 1 \text{ } \mu f$$

(answer)
$$R_x = R_4\frac{C_3}{C_s} = 10^3\left[\frac{10^{-11}}{10^{-9}}\right] = 10\Omega$$

If your answers are correct, go on to Problem 2-6.
If your answers are not correct, review p. 12.

PROBLEM 2-6. CIRCUIT NETWORK—LOOP CURRENT ANALYSIS

Find the current in the capacitor branch of the network shown below:

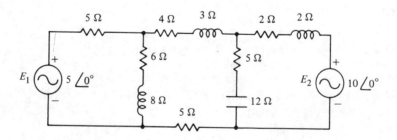

Solution:

Since there are three loops, three voltage equations may be written. The problem is simplified if the capacitor branch is included in only one loop, since the capacitor current is obtained directly. Therefore, write the loop equations as follows:

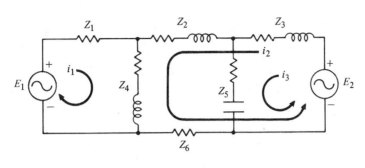

$$E_1 = i_1(Z_1 + Z_4) + i_2(Z_4) + i_3(0)$$

$$E_2 = i_1(Z_4) + i_2(Z_2 + Z_3 + Z_4 + Z_6) + i_3(Z_3)$$

$$E_2 = i_1(0) + i_2(Z_3) + i_3(Z_3 + Z_5)$$

Solving for i_3 using determinants:

$$\begin{bmatrix} E_1 \\ E_2 \\ E_2 \end{bmatrix} = \begin{bmatrix} (Z_1 + Z_4) & Z_4 & 0 \\ Z_4 & (Z_2 + Z_3 + Z_4 + Z_6) & Z_3 \\ 0 & Z_3 & Z_3 + Z_5 \end{bmatrix} \begin{bmatrix} i_1 \\ i_2 \\ i_3 \end{bmatrix}$$

$$i_3 = \cfrac{\begin{bmatrix} (Z_1 + Z_4) & Z_4 & E_1 \\ Z_4 & (Z_2 + Z_3 + Z_4 + Z_6) & E_2 \\ 0 & Z_3 & E_2 \end{bmatrix}}{\begin{bmatrix} Z_1 + Z_4 & Z_4 & 0 \\ Z_4 & (Z_2 + Z_3 + Z_4 + Z_6) & Z_3 \\ 0 & Z_3 & Z_3 + Z_5 \end{bmatrix}} = \cfrac{\begin{bmatrix} (11 + j8) & (6 + j8) & 5 \\ (6 + j8) & (17 + j13) & 10 \\ 0 & (2 + j2) & 10 \end{bmatrix}}{\begin{bmatrix} (11 + j8) & (6 + j8) & 0 \\ (6 + j8) & (17 + j13) & (2 + j2) \\ 0 & (2 + j2) & (7 - j10) \end{bmatrix}}$$

$$i_3 = \cfrac{10(11+j8)(17+j13) + 5(2+j2)(6+j8) - 10(6+j8)(6+j8) - 10(2+j2)(11+j8)}{(11+j8)(17+j13)(7-j10) - (6+j8)(6+j8)(7-j10) - (11+j8)(2+j2)(2+j2)}$$

$$= \cfrac{10(13.6\underline{/36°})(21.4\underline{/37.4°}) + 5(2.83\underline{/45°})(10\underline{/53.1°}) - 10(10\underline{/53.1°})(10\underline{/53.1°}) - 10(2.83\underline{/45°})(13.6\underline{/36°})}{(13.6\underline{/36°})(21.4\underline{/37.4°})(12.21\underline{/-55°}) - (10\underline{/53.1°})(10\underline{/53.1°})(12.21\underline{/-55°}) - (13.6\underline{/36°})(2.83\underline{/45°})(2.83\underline{/45°})}$$

$$= \cfrac{2910.4\ \underline{/73.4°} + 141.5\ \underline{/98.1°} - 1000\ \underline{/106.2°} - 384.88\ \underline{/81°}}{3553.6\ \underline{/18.4°} - 1221\ \underline{/51.2°} - 108.9\ \underline{/126°}}$$

$$= \cfrac{831.47 + j2789.1 - 19.94 + j140.09 + 278.99 - j960.29 - 60.21 - j380.14}{3371.93 + j1121.69 - 765.08 - j951.57 + 64.01 - j88.1}$$

$$= \cfrac{1030.31 + j1588.76}{2670.85 + j82.02} = \cfrac{1893.59\ \underline{/57.04°}}{2672.11\ \underline{/1.76°}} = 0.71\ \underline{/55.28°}\ \text{amp} \qquad \text{(answer)}$$

Alternate Solution:

Select three different loop equations:

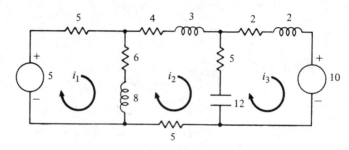

$$5 = i_1(11 + j8) - i_2(6 + j8) \tag{1}$$

$$0 = -i_1(6 + j8) + i_2(20 - j1) - i_3(5 - j12) \tag{2}$$

$$-10 = 0 - i_2(5 - j12) + i_3(7 - j10) \tag{3}$$

Using the first two equations to eliminate i_1 :

$$5(6 + j8) = i_1(11 + j8)(6 + j8) - i_2(6 + j8)(6 + j8)$$
$$\underline{\quad 0 = -i_1(11 + j8)(6 + j8) + i_2(20 - j1)(11 + j8) - i_3(5 - j12)(11 + j8)}$$
$$5(6 + j8) = i_2[(20 - j1)(11 + j8) - (6 + j8)(6 + j8)] - i_3(5 - j12)(11 + j8)$$

$$30 + j40 = i_2[(20.02\underline{/-2.86°})(13.6\underline{/36.03°}) - (10\underline{/53.1°})(10\underline{/53.1°})]$$
$$- i_3(13\underline{/-67.38°})(13.6\underline{/36.03°})$$

$$50\underline{/53.13°} = i_2[272.27\underline{/+33.17°} - 100\underline{/106.2°}] - i_3(176.8\underline{/-31.35°})$$

$$50\underline{/53.13°} = i_2(227.9 + j148.97 + 27.9 - j96.03) - i_3(176.8\underline{/-31.35°})$$

$$50\underline{/53.13°} = i_2(255.8 + j52.94) - i_3(176.8\underline{/-31.35°})$$

$$50\underline{/53.13°} = i_2(261.22\underline{/11.69°}) - i_3(176.8\underline{/-31.35°}) \tag{4}$$

Using equations (3) and (4) to eliminate i_2 :

$$(5 - j12)50\underline{/53.13°} = i_2(261.22\underline{/11.69°})(5 - j12) - i_3(176.8\underline{/-31.35°})(5 - j12)$$
$$-10(261.22\underline{/11.69°}) = -i_2(261.22\underline{/11.69°})(5 - j12) + i_3(7 - j10)(261.22\underline{/11.69°})$$

$$(13\underline{/-67.38°})(50\underline{/53.13°}) - 2612.2\underline{/11.69°}$$
$$= i_3[(261.22\underline{/11.69°})(12.21\underline{/-55.01°}) - (13\underline{/-67.38°})(176.8\underline{/-31.35°})]$$

$$650\underline{/-14.25} - 2612.2\underline{/11.69°} = i_3(3189.5\underline{/-43.32°} - 2298.4\underline{/-98.73°})$$

$$630 - j160 - 2558.02 - j529.27 = i_3(2320.47 - j2188.2 + 348.85 + j2271.77)$$

$$-1928.02 - j689.27 = i_3(2669.31 + j83.54)$$

$$2047.52\underline{/-160.33°} = i_3(2670.62\underline{/1.79°})$$

$$i_3 = \frac{2047.52\underline{/-160.33°}}{2670.62\underline{/1.79}} = 0.77\underline{/-162.12°}$$

Solving for i_2 using i_3 and equation (3) yields:

$$-10 = -i_2(13\underline{/-67.38°}) + (0.77\underline{/-162.12°})(12.21\underline{/-55.01°})$$

$$i_2 = \frac{10 + 9.4\underline{/-217.13°}}{13\underline{/-67.38°}} = -\frac{10 - 7.49 + j5.67}{13\underline{/-67.38°}} = \frac{2.51 + j5.67}{13\underline{/-67.38°}} = \frac{6.2\underline{/66.12°}}{13\underline{/-67.38°}}$$

$$= 0.48\underline{/133.5°}$$

Solving for capacitor current:

$$i_C = i_2 - i_3 = 0.48\underline{/133.5°} - 0.77\underline{/-162.12°}$$

$$= -0.33 + j0.35 + 0.73 + j0.24 = 0.4 + j0.58 = 0.71\underline{/55.62°}\text{amp} \quad \text{(answer)}$$

If your answer is correct, go on to Problem 2-7.
If your answer is not correct, review pp. 17, 20, and 21.

CAPACITORS

The problems in this section illustrate the fundamental properties of capacitor energy and charge.

PROBLEM 2-7. CAPACITOR ENERGY AND CHARGE

The current waveform shown below approximates a current impulse. If this pulse is applied to a capacitor, determine:
   a. how much energy is applied to the capacitor;
   b. how much charge it accepts.

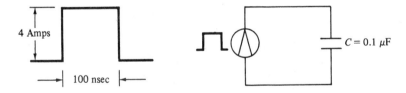

Solution:

Equations required for solution are:

$$W = \frac{1}{2} CV^2$$

$$Q = CV$$

$$V = \frac{1}{C} \int_0^T i \, dt \quad \text{(the integral represents the area under the current impulse curve)}$$

Solving for capacitor voltage:

$$V = \frac{1}{10^{-7}} \int_0^{10^{-7}} 4 \, dt = \frac{4 \times 10^{-7}}{10^{-7}} = 4 \text{ volts}$$

Therefore,

(answer) a. $W = \dfrac{10^{-7}}{2} (4)^2 = 8 \times 10^{-7}$ joule

(answer) b. $Q = 4 \times 10^{-7}$ coulomb

If your solution is correct, go on to Problem 2-8.
If your solution is not correct, review pp. 2, 3, 6 and 7.

PROBLEM 2-8.   TWO-CAPACITOR CHARGE TRANSFER

In the circuit shown below, the two capacitors are charged to the given initial conditions prior to switch closure. The value of the resistor is $0 < R < \infty$.

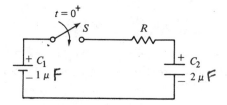

Initial conditions:

$$V_{C_1} = 100 \text{ volts} \qquad V_{C_2} = 25 \text{ volts}$$

Determine for the two capacitors the final values for:

a. voltage;
b. charge;
c. energy;
d. the energy dissipated as heat in the resistor after the switch has been closed for a long time.

Solution:

After the switch is closed, current will flow in the circuit until the two capacitors reach the same voltage (somewhere between 25 and 100). The length of time current flows is dependent, in part, upon the value of R. No matter what value R is, it will eventually dissipate a fixed amount of energy as heat. Since no energy is being added to the circuit by the principle of *conservation of charge*, the charge in the circuit will always remain unchanged; it will merely be redistributed after the switch is closed.

Initial Conditions:

$$Q_1 = C_1 V_1 = 10^{-6} \times 100 = 100 \times 10^{-6} \text{ coulomb}$$

$$Q_2 = C_2 V_2 = 2 \times 10^{-6} \times 25 = 50 \times 10^{-6} \text{ coulomb}$$

$$Q_{\text{tot}} = Q_1 + Q_2 = 150 \times 10^{-6} \text{ coulomb}$$

$$W_1 = \tfrac{1}{2} C_1 V_1^2 = \tfrac{1}{2} \times 10^{-6} \times 10^4 = 5 \times 10^{-3} \text{ joule}$$

$$W_2 = \tfrac{1}{2} C_2 V_2 = \tfrac{1}{2} \times 2 \times 10^{-6} \times 625 = 0.625 \times 10^{-3} \text{ joule}$$

$$W_{\text{tot}} = W_1 + W_2 = 5.625 \times 10^{-3} \text{ joule}$$

Final Conditions:

$$Q_{\text{tot}} = 150 \times 10^{-6} \text{ coulomb} = Q_1 + Q_2 = C_1 V_1 + C_2 V_2 = (C_1 + C_2)V$$

due to conservation of charge and $V_1 = V_2$

a. $V = \dfrac{Q_{\text{tot}}}{C_1 + C_2} = \dfrac{150 \times 10^{-6}}{3 \times 10^{-6}} = 50 \text{ volts}$ (answer)

b. $Q_1 = 50C_1 = 50 \times 10^{-6} \text{ coulomb}$ (answer)

    $Q_2 = 50C_2 = 100 \times 10^{-6} \text{ coulomb}$

c. $W_1 = \tfrac{1}{2} C_1 V^2 = \tfrac{1}{2} \times 10^{-6} \times 2500 = 1.25 \times 10^{-3} \text{ joule}$ (answer)

    $W_2 = \tfrac{1}{2} C_2 V^2 = \tfrac{1}{2} \times 2 \times 10^{-6} \times 2500 = 2.5 \times 10^{-3} \text{ joule}$

    $W_{\text{tot}} = W_1 + W_2 = 3.75 \times 10^{-3} \text{ joule}$

d. Energy lost in the resistor is the difference between the initial total capacitor energy and the final total capacitor energy. Therefore:

$$W_R = (5.625 - 3.75)\, 10^{-3} = 1.875 \times 10^{-3} \text{ joule} \qquad \text{(answer)}$$

If your solutions are correct, go on to Problem 2-9.
If your solutions are not correct, review pp. 2, 3, 6, 7, 10, 15 and 16.

PROBLEM 2-9. POWER, ENERGY, AND CHARGE

In the steady-state circuit below, calculate:

  a. voltage across the capacitor;
  b. stored charge;
  c. energy;
  d. current delivered by the battery;
  e. power delivered by the battery.

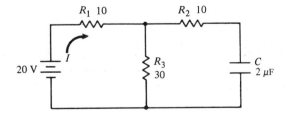

Solution:

Since conditions are steady-state, there is no current flow in the capacitor loop. Therefore, resistors R_1 and R_3 act as a voltage divider and:

(answer)   a. $V = 20 \dfrac{R_3}{R_1 + R_3} = 20 \dfrac{30}{40} = 15$ volts

(answer)   b. $Q = CV = 30 \times 10^{-6}$ coulomb

(answer)   c. $W = \frac{1}{2} CV^2 = 225 \times 10^{-6}$ joule

(answer)   d. $I = \dfrac{V}{R} = \dfrac{20}{40} = 0.5$ amp

(answer)   e. $P = I^2 R = (0.5)^2\, 40 = 10$ watts

If your solutions are correct, go on to Problem 2-10.
If your solutions are not correct, review pp. 2, 3, 6 and 7.

AC CIRCUITS

The problems in this section illustrate that Ohm's Law works for ac circuits as well as dc circuits.

PROBLEM 2-10.   UNKNOWN DEVICE

You have been given a potted device and asked to determine the type and value of elements it contains. When connected to 120 volts, 60 Hz, 2 amps ac flow in each of the three paths, shown below:

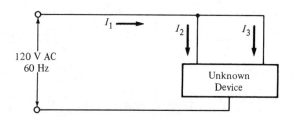

Assume all elements are linear, passive, and bidirectional.

Solution:

In order for all three currents to be the same magnitude, the current phasor diagrams must form an equilateral triangle as shown below (assume that I_1 is at $0°$ with respect to the ac source and that branch 2 is capacitive).

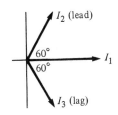

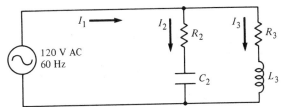

$$I_1 = I_2 \underline{/+60^\circ} + I_3 \underline{/-60^\circ} = 2 \underline{/0^\circ} \text{ amps}$$

$$Z_2 = \frac{120 \underline{/0^\circ}}{2 \underline{/+60^\circ}} = 60 \underline{/-60^\circ} = 30 - j51.96$$

$$R_2 = 30 \ \Omega$$

$$X_{C_2} = 51.96 \ \Omega \qquad \text{(answer)}$$

$$C_2 = \frac{1}{2\pi f X_{C_2}} = \frac{1}{377 \times 51.96} = 5.11 \ \mu f$$

$$Z_3 = \frac{120 \underline{/0^\circ}}{2 \underline{/-60^\circ}} = 60 \underline{/60^\circ} = 30 + j51.96 \qquad \text{(answer)}$$

$$R_3 = 30 \ \Omega$$

$$X_{L_3} = 51.96 \ \Omega \qquad \text{(answer)}$$

$$L_3 = \frac{X_{L_3}}{2\pi f} = \frac{51.96}{377} = 0.14 \ h \qquad \text{(answer)}$$

If your answers are correct, go on to Problem 2-11.
If your answers are not correct, review pp. 11 and 12.

PROBLEM 2-11. PARALLEL BRANCHES

In the circuit that follows, calculate the power dissipated by the load:

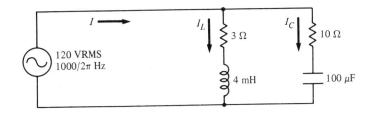

Solution:

$$X_L = 2\pi f L = 2\pi \times \frac{1000}{2\pi} \times 4 \times 10^{-3} = 4 \ \Omega$$

41

$$X_C = \frac{1}{2\pi f C} = \frac{1}{2\pi \times \dfrac{1000}{2\pi} \times 100 \times 10^{-6}} = 10 \ \Omega$$

$$Z_L = 3 + j4 = 5 \underline{/53.13°} \ \Omega$$

$$Z_C = 10 - j10 = 14.14 \underline{/-45°} \ \Omega$$

$$I_C = \frac{E}{Z_C} = \frac{120 \underline{/0°}}{14.14 \underline{/-45°}} = 8.49 \underline{/45°} = 6 + j6 \text{ amps}$$

$$I_L = \frac{E}{Z_L} = \frac{120 \underline{/0°}}{5 \underline{/53.13°}} = 24 \underline{/-53.13°} = 14.4 - j19.2 \text{ amps}$$

$$I = I_L + I_C = 20.4 - j13.2 = 24.3 \underline{/-32.91°}$$

(answer)
$$P = VI \cos \theta = 120 \times 24.3 \cos 32.91° = 2448 \text{ watts}$$

If your answer is correct, go on to Problem 2-12.
If your answer is not correct, review pp. 11 and 12.

COMPENSATING CIRCUITS

The problems in this section apply to a variety of electrical engineering specialities including control theory and transient analysis. Use is made of partial fractions (see Chapter 5) and the Laplace transform.

PROBLEM 2-12. SIMPLE LAG CIRCUIT

Derive the equation for output voltage across the capacitor of the simple lag circuit shown below after the switch is closed. Assume the capacitor is initially uncharged.

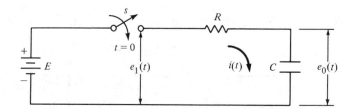

Solution:

$$e_1(t) = Ri(t) + \frac{1}{C} \int_0^t i \, dt \xrightarrow{\mathcal{L}} E_1(s) = I(s) \left[R + \frac{1}{Cs} \right]$$

$$e_0(t) = \frac{1}{C} \int_0^t i \, dt \xrightarrow{\mathcal{L}} E_0(s) = \frac{I(s)}{Cs}$$

42

The transfer function of a circuit $= \dfrac{\text{output}}{\text{input}} = $ T.F.

$$\frac{E_0(s)}{E_1(s)} = \left[\frac{I(s)}{Cs}\right]\left[\frac{1}{I(s)(R+1/Cs)}\right] = \frac{1}{RCs+1} = \frac{1/RC}{s+1/RC}$$

When S is closed at $t = 0$, a transient unit step is input.

$$\mathcal{L}e_1(t) = e_1(s) = \frac{E}{s}$$

$$E_0(s) = E_1(s) \times \text{T.F.} = \left[\frac{E}{s}\right]\left[\frac{1/RC}{s+1/RC}\right]$$

Using partial fractions,

$$\left[\frac{E}{s}\right]\left[\frac{1/RC}{s+1/RC}\right] = \frac{AE}{s} + \frac{B/RC}{s+1/RC} = \frac{AE(s+1/RC)+Bs/RC}{s(s+1/RC)}$$

Cancelling common denominators,

$$\frac{E}{RC} = AE(s+1/RC) + \frac{Bs}{RC}$$

Solving for A and B:

$$\text{Let } s = 0, \frac{E}{RC} = \frac{AE}{RC}, A = 1$$

$$\text{Let } s = -\frac{1}{RC}, \frac{E}{RC} = -\frac{B/RC}{RC}, B = -ERC$$

Substituting:

$$E_0(s) = \frac{AE}{s} + \frac{B/RC}{s+1/RC} = \frac{E}{s} - \frac{ERC/RC}{s+1/RC} = E\left[\frac{1}{s} - \frac{1}{s+1/RC}\right]$$

Taking the inverse $\mathcal{L}$,

$$e_0(t) = E(1 - e^{-t/RC}) \qquad \text{(answer)}$$

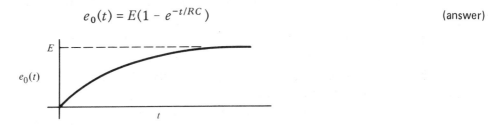

If your answer is correct, go on to Problem 2-13.
If your answer is not correct, review pp. 16, 61 and 25.

PROBLEM 2-13. SIMPLE LEAD CIRCUIT

What is the equation for output voltage across the resistor of the following simple lead circuit after the switch is closed? Assume the capacitor is initially uncharged.

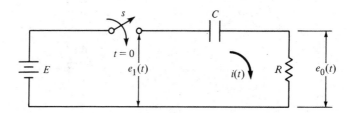

Solution:

The derivation of the voltage response across the resistor is similar to that of the preceding problem. Therefore, some short cuts are taken here.

Using Kirchhoff's voltage law,

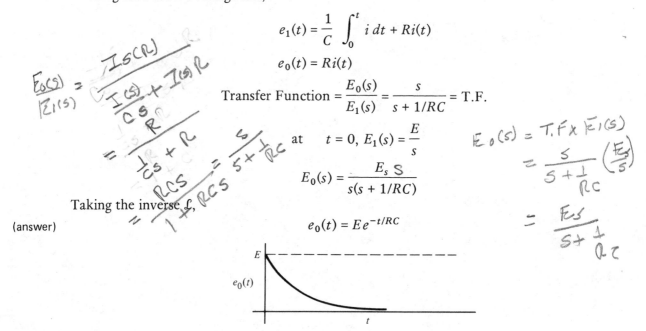

$$e_1(t) = \frac{1}{C} \int_0^t i\, dt + Ri(t)$$

$$e_0(t) = Ri(t)$$

$$\text{Transfer Function} = \frac{E_0(s)}{E_1(s)} = \frac{s}{s + 1/RC} = \text{T.F.}$$

at $\quad t = 0, E_1(s) = \dfrac{E}{s}$

$$E_0(s) = \frac{E_s\, S}{s(s + 1/RC)}$$

Taking the inverse $\mathcal{L}$,

(answer)

$$e_0(t) = E e^{-t/RC}$$

If your answer is correct, go on to Problem 2-14.
If your answer is not correct, review the solution to Problem 2-12.

VOLTMETERS

This problem illustrates the practical application of using a microammeter and some precision resistors to create a voltmeter.

PROBLEM 2-14. VOLTMETER DESIGN

Given a microammeter having a 50 μ amp movement and an internal resistance of 10 ohms, design a voltmeter with ranges of 10 VDC, 50 VDC, 100 VDC, and 500 VDC.

Solution:

The voltmeter circuit is as follows:

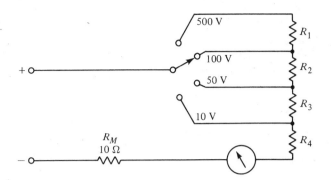

A 50 μ amp movement yields a sensitivity of:

$$\frac{1}{50 \times 10^{-6}} = 20,000 \text{ ohms/volt}$$

$$I = \frac{V}{R} \qquad \frac{1}{I} = \frac{R}{V}$$

Calculating the precision resistors:

$$\frac{10 V}{R_4 + 10\Omega} = 50\times10^{-6} \qquad \frac{10}{50\times10^{-6}} = R_4 + 10$$

Resistance of the 10 volt range, R_{10}, is:

$$R_{10} = R_4 + R_M = (10 \text{ volts}) (20,000 \text{ }\Omega/\text{volt}) = 200,000 \text{ }\Omega$$

$$R_4 = 200,000 - 10 = 199,990 \text{ }\Omega$$

$$.2\times10^{+6} = 2\times10^{5} = R_4 + 10$$
$$R_4 = 200,000$$
(answer) $$-10$$
$$= 199,990$$

Resistance of the 50 volt range, R_{50}, is:

$$R_{50} = R_3 + R_{10} = (50 \text{ volts}) (20,000 \text{ }\Omega/\text{volt}) = 1 \text{ M}\Omega$$

$$R_3 = 10^6 - 20,000 = 800,000 \text{ }\Omega$$

$$50\times10^{-6} = \frac{50 V}{2\times10^{5} + R_3}$$
(answer)

$$2\times10^{5} + R_3 = 10\times10^{5}$$
$$= 8\times10^{5}$$

Resistance of the 100 volt range, R_{100}, is:

$$R_{100} = R_2 + R_{50} = (100 \text{ volts}) (20,000 \text{ }\Omega/\text{volt}) = 2 \text{ M}\Omega$$

$$R_2 = 2 \times 10^6 - 10^6 = 1 \text{ M}\Omega$$
(answer)

Resistance of the 500 volt range, R_{500}, is:

$$R_{500} = R_1 + R_{100} = (500 \text{ volts}) (20,000 \text{ }\Omega/\text{volt}) = 10 \text{ M}\Omega$$

$$R_1 = 10 \times 10^6 - 2 \times 10^6 = 8 \text{ M}\Omega$$
(answer)

If your answers are correct, go on to Problem 2-15.
If your answers are not correct, recheck your calculations.

IMPEDANCE TRANSFORMATION

There are several methods for matching circuits by means of impedance transformation. Chapter 6 illustrates the use of a stub in parallel with a high-frequency transmission line. The problem in this section illustrates an in-line method using an L-section. Ryder[17] gives many additional examples, including T-section, exponential line, T-network, and tapped resonant circuit.

PROBLEM 2-15.   IMPEDANCE TRANSFORMATION

For the circuit shown below, find the load-section parameters in order to match the load to the generator for maximum power transfer.

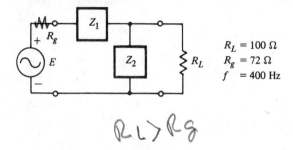

$R_L = 100 \ \Omega$
$R_g = 72 \ \Omega$
$f = 400$ Hz

Solution:

$$R_L > R_g$$

Refer to the section on resonance in Chapter 1. This problem fits the case in which $R_L > R_g$. Therefore, Z_1 is an inductor L and Z_2 is a capacitor C the values of which are calculated as follows:

(answers)

$$L = \frac{R_g}{\omega} \sqrt{\frac{R_l}{R_g} - 1} = \frac{72}{2\pi \times 400} \sqrt{\frac{100}{72} - 1} = 0.0179 \text{ h}$$

$$C = \frac{1}{\omega R_l} \sqrt{\frac{R_l}{R_g} - 1} = \frac{1}{2\pi \times 400 \times 100} \sqrt{\frac{100}{72} - 1} = 2.48 \ \mu\text{f}$$

If your answers are correct, go on to Problem 2-16.
If your answers are not correct, review p. 26.

AMMETERS

The problem in this section illustrates the difference in ac and dc current measurement.

PROBLEM 2-16.   AC AND DC AMMETERS

In the circuit below an ac and a dc source are connected in series with a 100 ohm load. Ammeters A_1 and A_2 are also in series with the circuit to measure ac and dc current. A_1 is a true rms ammeter, and A_2 is a D'Arsonval-type dc ammeter.
   Determine:

   a. the reading of A_1;
   b. the reading of A_2;
   c. the power dissipated in the load.

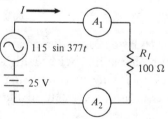

Solution:

Ac voltmeters and ammeters for power frequencies are designed to indicate effective value, even for nonsinusoidal waveforms. Instruments for higher frequencies also give the effective value for sine waves, but for other waveforms, the reading is not the effective reading. The composite current I has an ac component and a dc component. These two components are calculated as follows:

$$I_{ac} = \frac{115}{100\sqrt{2}} = 0.813 \text{ amp}$$

$$I_{dc} = \frac{25}{100} = 0.25 \text{ amp}$$

a. Meter A_1 reads I_{RMS}, which is equal to the square root of the sum of the squares of the current components:

$$A_1 = I_{RMS} = \sqrt{I_{ac}^2 + I_{dc}^2} = \sqrt{0.813^2 + 0.25^2}$$
$$= \sqrt{0.66 + 0.0625} = \sqrt{0.724} = 0.85 \text{ amp} \qquad \text{(answer)}$$

b. Meter A_2 reads $I_{dc} = 0.25$ amp. (answer)

c. Power dissipated in the load is: (answer)

$$P_L = I_{RMS}^2 R_L = 0.85^2 \times 100 = 72.4 \text{ watts}$$

If your answers are correct, go to Problem 2-17.
If your answers are not correct, review p. 27.

RESONANCE

The subject of resonance is quite broad. This section contains a variety of problems illustrating this subject. Ryder[17] is a good source for further study in this area.

PROBLEM 2-17. SERIES RESONANCE

A series resonant circuit as shown below has a resonant frequency of 3000 Hz. Find:

a. coil inductance L;
b. maximum value of coil resistance R if the bandwidth is not to exceed 100 Hz.

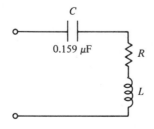

Solution:

a. At resonant frequency, $\omega L = 1/\omega C$

(answer)
$$L = \frac{1}{\omega^2 C} = \frac{1}{(2\pi \times 3000)^2 \times 0.159 \times 10^{-6}} = 0.017 \text{ h}$$

b. Bandwidth is $B = f_r/Q = 100$ Hz

$$Q = \frac{3000}{100} = 30$$

$$Q = \frac{1}{R} \sqrt{\frac{L}{C}}$$

Therefore,

(answer)
$$R = \frac{1}{Q} \sqrt{\frac{L}{C}} = \frac{1}{30} \sqrt{\frac{0.0177}{0.159 \times 10^{-6}}} = 11.12 \ \Omega$$

If your answers are correct, go on to Problem 2-18.
If your answers are not correct, review p. 26.

PROBLEM 2-18. PASSIVE CIRCUIT CALCULATION

Compute the inductance of a coil required to produce parallel resonance at 60 Hz in a circuit with a 50 KVAR-rated capacitor at 4800 volts.

Solution:

$$\frac{50,000 \text{ VAR}}{4,800 \text{ V}} = 10.42 \text{ amps}$$

$$X_C = \frac{4,800}{10.42} = 461 \ \Omega$$

For parallel resonance: $|B_L| = |B_C|$

For the special case of non-dissipative circuits:

$$\frac{1}{\omega L} = \omega C, \text{ or } |X_L| = |X_C|$$

Therefore, $|X_L| = |X_C| = 461 \ \Omega$

(answer)
$$L = \frac{X_L}{\omega} = \frac{461}{377} = 1.22 \text{ h}$$

If your answer is correct, go on to Problem 2-19.
If your answer is not correct, review pp. 12, 13 and 23.

PROBLEM 2-19. SERIES-PARALLEL RESONANT FILTER

A filter is a device which passes currents of a certain frequency and offers high impedance to currents of another frequency. Many complex reactive networks have been developed as active filters over the years. (They are described in detail in textbooks and handbooks[7,36].)

The circuit below is required to pass 60 KHz with minimum impedance and must block 30 KHz current as effectively as possible.

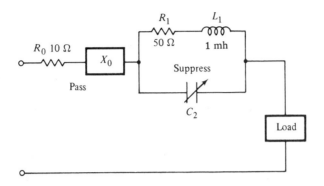

Solve for C_2 and X_0 to obtain the stated requirements.

Solution:

This filter is designed in two steps: first determine C_2 in the parallel antiresonant circuit for the suppress frequency; then solve for the reactance required for pass frequency in the series resonant branch.

1. Calculate C_2 to produce parallel resonance:

$$X_1 = 2\pi f L_1 = 2\pi \times 30 \times 10^3 \times 10^{-3} = 188.5\,\Omega$$

$$Z_1 = R_1 + jX_1 = 50 + j188.5 = 195\underline{/75.14°}$$

$$Y_1 = \frac{1}{Z_1} = 0.0051278\underline{/-75.14°}\ \text{mho} = 0.001315 - j0.0049563$$

$$= \quad G_1 \quad - \quad jB_1$$

For parallel resonance, $|B_{C_2}| = |B_1|$

$$B_{C_2} = 0.0049563\ \text{mho} = 2\pi f C_2\ \mho$$

$$X_{C_2} = \frac{1}{B_{C_2}} = 201.762\ \Omega$$

$$C_2 = \frac{0.0049563}{2\pi \times 30 \times 10^3} = 2.6294 \times 10^{-8} = 0.026294\ \mu f \qquad \text{(answer)}$$

The parallel branch now has an admittance Y_P and impedance Z at 30 KHz if:

$$Y_P = 0.001315\ \text{mho}$$

$$Z_P = 760.4\underline{/0°}\ \Omega$$

2. Calculate X_0 and its capacitance or inductance to produce series resonance. Recalculating the parallel circuit parameters at the pass frequency of 60 KHz:

$$X_L = 2\pi f L = 2\pi \times 60 \times 10^3 \times 10^{-3} = 377\ \Omega$$

$$X_C = \frac{1}{2\pi f C_2} = \frac{1}{2\pi \times 60 \times 10^3 \times 2.6294 \times 10^{-8}} = 100.88\ \Omega$$

The new parallel impedance becomes:

$$Z_P = \frac{(50 + j377)(-j100.88)}{50 + j377 - j100.88} = \frac{(380.3\underline{/82.45°})(100.88\underline{/-90°}}{50 + j276.12}$$

$$= \frac{38364.78\underline{/-7.55°}}{280.61\underline{/79.74°}} = 136.72\underline{/-87.29°} = 6.45 - j136.57$$

Since Z_P is capacitive at 60 KHz, X_0 must be inductive to make the total circuit purely resistive.

$$X_0 = j136.57 = 2\pi f L_0$$

(answer)
$$L_0 = \frac{136.57}{2\pi \times 60 \times 10^3} = 0.362\ \text{mh}$$

If your answer is correct, go to Problem 2-20.
If your answer is not correct, review pp. 12, 13 and 23.

PROBLEM 2-20.   SERIES RESONANT FILTER

A small company is experiencing an annoying 780 Hz tone on its telephone circuits. It has been determined that the tone is due to power and telephone lines running parallel for a long distance. For some reason, the thirteenth harmonic of the power line is being induced into the telephone circuits at an intolerable level. To correct the problem, a three-phase delta-connected shunt will be installed on the power line that will present 15 ±1% ohms to the harmonic and 10,000 ±1% ohms to the fundamental 60 Hz frequency.

Determine the parametric values of the series-resonant shunt that will meet these specifications.

Solution:

The schematic for one phase of the shunt is as follows:

At 780 Hz, the lowest impedance is desired, which is achieved by series resonance. Therefore,

$$X_L = X_C \text{ and } R = 15\ \Omega$$

At 60 Hz, high impedance is desired. At low frequencies, X_C offers the highest impedance. Therefore, X_C should be close to the desired 10,000 ohms. We can try a value of 10,060 ohms for X_C at 60 Hz (X_L will subtract from this to bring it closer to 10,000 ohms) and see if the result is within the 1% tolerance.

$$\text{At 60 Hz, } C = \frac{1}{2\pi f X_C} = \frac{1}{377 \times 10,060} = 0.264 \ \mu\text{f}$$

$$\text{At 780 Hz, } X_C = \frac{1}{2\pi f C} = \frac{1}{13 \times 377 \times 0.264 \times 10^{-6}} = 774 \ \Omega$$

For series resonance: $|X_L| = |X_C| = 774$

$$L = \frac{X_L}{2\pi f} = \frac{774}{13 \times 377} = 0.158 \ \text{h}$$

Checking at 60 Hz:

$$X_L = 2\pi f L = 377 \times 0.158 = 59.5 \ \Omega$$

$$Z = R + jX_L - jX_C = 15 + j59.5 - j10,060$$

$$= 15 - j10,000 = 10,000 \underline{/-89.9°} \ \Omega$$

This is well within the permitted tolerance. To summarize, the values of R, L, and C are:

$R = 15 \ \Omega$ (answer)
$L = 0.158 \ \text{h}$ (answer)
$C = 0.264 \ \mu\text{f}$ (answer)

If your answers are correct, go on to Problem 2-21.
If your answers are not correct, review the solution to Problem 2-18.

PROBLEM 2-21. RLC BANDPASS FILTER

A series RLC circuit is driven by a fixed-amplitude swept-frequency generator. It is desired that the current be maximum at 1500 Hz and down 3 db at 1400 and 1600 Hz. Determine the Q of the circuit and the values of R and C if $L = 50$ mh.

Solution:

At the series resonant frequency (where maximum current exists):

$$X_L = 2\pi f L = 2\pi \times 1500 \times 50 \times 10^{-3} = 471 \ \Omega$$

$$|X_C| = |X_L| = 471 = \frac{1}{2\pi f C}$$

$$C = \frac{1}{2\pi f X_C} = \frac{1}{2\pi \times 1500 \times 471} = 0.225 \ \mu\text{f} \qquad \text{(answer)}$$

$$Q = \frac{f_r}{B} = \frac{1500}{200} = 7.5 \qquad \text{(answer)}$$

(answer)

$$Q = \frac{1}{R}\sqrt{\frac{L}{C}}, \quad R = \frac{1}{Q}\sqrt{\frac{L}{C}} = \frac{1}{7.5}\sqrt{\frac{50 \times 10^{-3}}{0.225 \times 10^{-6}}} = 62.85 \; \Omega$$

If your answers are correct, go on to Problem 2-22.
If your answers are not correct, review p. 24.

PROBLEM 2-22. DAMPED RLC CIRCUIT

The following circuit represents a system in which a dc voltage is periodically applied, causing a current pulse which never goes negative. The capacitor is discharged after each cycle. If the values of L and C are 20 mh and 10 μf, what is the minimum value of R in order to achieve the desired result?

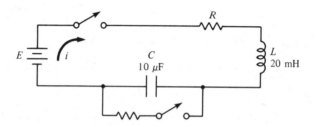

Solution:

The desired current waveform for one period is:

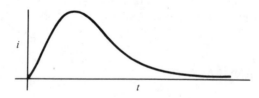

The circuit must be critically damped if R is to be a minimum while the current is always positive.

Applying Kirchhoff's voltage law:

$$E(t) = iR + L\frac{di}{dt} + \frac{1}{C}\int i \, dt$$

In Laplace form:

$$E(s) = I(s)\left[R + sL + \frac{1}{sC}\right]$$

The characteristic equation is:

$$R + sL + \frac{1}{sC} = 0$$

or

$$s^2L + sR + \frac{1}{C} = 0$$

Applying the quadratic equation to solve for the roots of the equation:

$$s = -\frac{R}{2L} \pm \sqrt{\frac{R^2}{4L^2} - \frac{1}{LC}}$$

This is of the form:

$$-\alpha \pm \sqrt{\alpha^2 - \omega_r{}^2}$$

where

$$\omega_r = \frac{1}{\sqrt{LC}} \text{ (resonant frequency)}$$

$$\alpha = \frac{R}{2L} \text{ (attenuation)}$$

For critical damping, the expression under the radical must equal zero.

Therefore,

$$\alpha^2 = \omega_r{}^2$$

or

$$\frac{R}{2L} = \frac{1}{\sqrt{LC}}$$

Solving for R:

$$R = \frac{2L}{\sqrt{LC}} = 2\sqrt{\frac{L}{C}} = 2\sqrt{\frac{20 \times 10^{-3}}{10 \times 10^{-6}}} = 89.4\ \Omega \qquad \text{(answer)}$$

If your answer is correct, go on to Problem 2-23.
If your answer is not correct, review p. 24.

PROBLEM 2-23. SHUNT PEAKING

One stage of a video amplifier is used to drive a 3.3 K ohm load. Since a wider bandwidth is required, it is decided to use shunt peaking compensation. Output resistance of the amplifier is 1 M ohm, and the equivalent output shunt capacitance is 27 pf.

Determine the value of L required to provide a normalized gain of 1 at the upper 3 db uncompensated frequency. Then determine the overshoot and the new 3 db bandwidth. Assume flat response to 0 Hz.

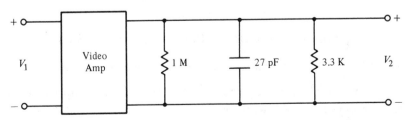

Solution:

The new circuit is as follows:

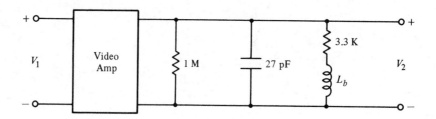

Old and new frequency response curves are compared below. Martin[19] has an excellent section on shunt peaking, and certain of his data are referenced in the solution of this problem.

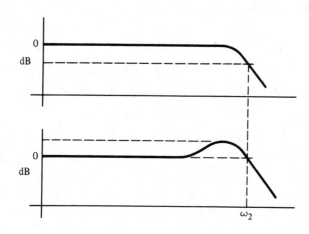

$$RC_T = \frac{1}{\omega_2}$$

where

ω_2 = upper cutoff frequency without L
$C_T = 27$ pf
$R = 1\,M \,//\, 3.3\,K = R_2 \,//\, R_L$

$$m = \frac{\omega_2 L_b}{R_L} = \text{peaking parameter} = \frac{L_b}{R_L{}^2 C_T}$$

$$L_b = m R_L R C_T = m \times (3.3 \times 10^3)^2 \times 27 \times 10^{-12}$$

A practical range for m is 0.3 to 0.4 (Martin[19], Figure 4.22). No peaking occurs at less than 0.25, and above 0.6 overshoot becomes excessive. Therefore, a value for m of 0.4 will be used here.

(answer)
$$L_b = 0.4 \times (3.3 \times 10^3)^2 \times 27 \times 10^{-12} = 0.118 \text{ mh}$$

(answer) From Figure 4.23 of Martin[19], it is seen that for $m = 0.4$, overshoot is 2.5%.

(answer) From Figure 4.24 of Martin[19], it is seen that for $m = 0.4$, the new upper cutoff frequency is $1.7\omega_2$. Therefore, the new 3 db bandwidth is 1.7 times the original bandwidth.

If your answers are correct, go on to Problem 2-24.

If your answers are not correct, review the section in Martin[19], or other appropriate texts, on shunt peaking compensation.

MAXIMUM POWER

Chapter 1 defines the maximum power transfer theorem and its corollary. The problems in this section illustrate these principles.

PROBLEM 2-24.   THEVENIN CIRCUIT—MAXIMUM POWER

The circuit below has properties found in electronic circuits. Convert the circuit to a Thevenin equivalent and determine the power dissipated in the load resistance when it is adjusted for maximum power transfer.

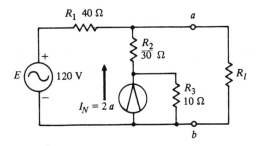

Solution:

Convert the Norton circuit, consisting of R_3 and the current source, to a Thevenin equivalent:

$$E_T = I_N R_3 = 20 \text{ volts}$$

$$R_T = R_3 = 10 \ \Omega$$

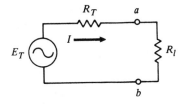

$$\frac{100 V}{80 \Omega} = 1.25 A$$

$$\frac{40}{50.00}$$

$$= 70 V$$

Now convert the circuit to the left of terminals ab to a Thevenin equivalent:

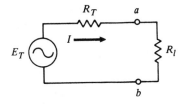

E_T = voltage at terminal a when R_l is not connected. Using superposition:

(answer)
$$E_T = 120\left(\frac{40}{80}\right) + 20\left(\frac{40}{80}\right) = 70 \text{ volts}$$

(answer)
$$R_T = \frac{(40)(40)}{40 + 40} = 20\ \Omega$$

For maximum power, $R_L = R_T = 20\ \Omega$; thus:

(answer)
$$P_l = I^2 R_l = \left(\frac{E_T}{R_T + R_l}\right)^2 R_l = \left(\frac{70}{40}\right)^2 20 = 61.25 \text{ watts}$$

If your answers are correct, go to Problem 2.25.
If your answers are not correct, review pp. 18 and 19.

PROBLEM 2-25. IDEAL TRANSFORMER—MAXIMUM POWER

Power is supplied by a generator to a transformer-coupled load as shown below. The internal impedance of the generator is $50 + j20$ ohms, and the frequency is 400 Hz. Open circuit emf of the generator is 100 VRMS. What should be the turns ratio of the transformer and the value of the tuning capacitor to achieve maximum power transfer?

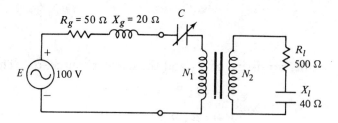

Solution:

To achieve maximum power, R_l must equal R_g and the reactances must cancel out. The equivalent circuit referred to the transformer primary is:

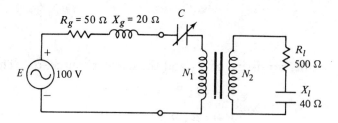

Impedances are proportional to the square of the turns ratio. Therefore:

(answer)
$$\left(\frac{N_1}{N_2}\right)^2 = \frac{R_g}{R_l} = \frac{50}{500} = \frac{1}{10}, \quad \frac{N_1}{N_2} = \sqrt{\frac{1}{10}}$$

Calculating the value for the variable capacitor:

$X_d = 16\, \Omega.$ $C = \dfrac{1}{2\pi f X_C} = \dfrac{1}{2\pi \times 400 \times 16} = 25\ \mu f$ (answer)

If your answers are correct, go to Problem 2-26.
If your answers are not correct, review pp. 17, 26 and 27.

PROBLEM 2-26.   MAXIMUM POWER TRANSFER

A network contains dc generators and resistors and has only two output terminals available for measurement. When the output terminals are shorted, six amps flow. When a 12 ohm resistor is connected between the two terminals, three amps flow. Determine the maximum power that may be taken from the network at the two output terminals.

Solution:

The following circuit may be used to represent the network:

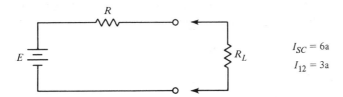

Writing two equations:

(1) $R = \dfrac{E}{6},$ $E = 6R$

(2) $R + 12 = \dfrac{E}{3},$ $E = 3R + 36$

(1) – (2) $0 = 3R - 36$

$R = 12\ \Omega$

$E = 6R = 72$ volts

For maximum power transfer, $R = R_L$

$$I = \frac{E}{R + R_L} = \frac{72}{24} = 3 \text{ amps}$$

$$P_L = I^2 R_L = 9 \times 12 = 108 \text{ watts}$$ (answer)

If your answer is correct, go on to Problem 2-27.
If your answer is not correct, review p. 18.

PROBLEM 2-27.   MAXIMUM POWER TRANSFER COROLLARY

Power at 1200 Hz is supplied to a load through a transformer by a signal generator having an internal impedance of $520 + j300$. It is desired to transfer maximum power to a resistive load of 20 ohms through a step-down transformer. Open circuit voltage output of the generator is 75 volts RMS.

Calculate the transformer turns ratio and power supplied to the load.

Solution:

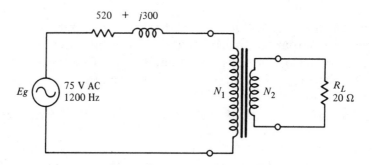

Since the resistance cannot be cancelled, then maximum power transfer occurs when the equivalent load impedance equals the generator impedance.

$$Z_g = 520 + j300 = 600\underline{/30°}.$$

$$|Z_g| = |a^2 R_L|$$

Assuming an ideal transformer (no losses):

(answer)
$$a = \sqrt{\frac{Z_g}{R_L}} = \sqrt{\frac{600}{20}} = \sqrt{30}$$

Equivalent R_L referred to the primary is 600 Ω.

$$I = \frac{Eg}{Z_g + R_L'} = \frac{75}{520 + 600 + j300} = \frac{75}{1159.5} = 64.7 \text{ ma}$$

(answer)
$$P_L = I^2 R_L = (64.7 \times 10^{-3})^2 600 = 2.51 \text{ watts}$$

If your answers are correct, go on to the next section.
If your answers are not correct, review pp. 19 and 26.

TRANSIENTS

The study of transients is diverse and applies to nearly all aspects of engineering. Chapter 1 contains a brief summary of the subject. Gardner and Barnes[11] is a classic reference. Problems in this section illustrate a few aspects of this subject that are likely to apply to the PE exam.

PROBLEM 2-28. RL TRANSIENT

In the circuit shown below, assume that steady-state conditions exist prior to switch closure. Determine $i_2(t)$ for the period after the switch is closed.

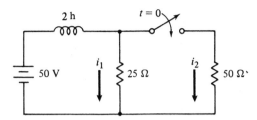

Solution:

This problem may be solved using the general transient response formula:

$$i(t) = i_{ss} + (i_0 - i_{ss})e^{-t/\tau}$$

Immediately after the switch is closed, there is no current change through the coil. However, one-third of the current flows immediately in the second branch; the other two-thirds flow through branch 1. Initial current in branch 2 is:

$$i_0 = \frac{50}{25} \times \frac{1}{3} = \frac{2}{3} \text{ amp}$$

Final current flow in branch 2 is:

$$i_{ss} = \frac{50}{\dfrac{25 \times 50}{75}} \times \frac{1}{3} = 1 \text{ amp}$$

The time constant for the transient is:

$$\tau = \frac{L}{R} = \frac{2}{16.67} = 0.12 \text{ seconds} \qquad \text{(answer)}$$

Now the transient response formula may be specified:

$$i_2(t) = 1 - \tfrac{1}{3}e^{-t/0.12}$$

If your answer is correct, go to Problem 2-29.
If your answer is not correct, review p. 13.

PROBLEM 2-29. DOUBLE ENERGY TRANSIENT

In the circuit shown below, assume steady-state conditions exist prior to switch opening. Determine $V_C(t)$ for capacitor voltage after the switch is opened.

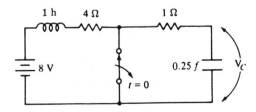

Solution:

This problem is best solved by means of Laplace transforms.

Initial conditions:

$$V_C = 0$$

$$I_0 = \frac{8}{4} = 2 \text{ amps}$$

After switch is opened:

$$E = i(t)R + L\frac{di}{dt} + \frac{1}{C}\int_0^t i\,dt$$

$$\frac{E}{s} = RI(s) + L[sI(s) - I_0] + \frac{1}{Cs}I(s)$$

$$I(s)\left[R + sL + \frac{1}{Cs}\right] = \frac{E}{s} + LI_0$$

$$I(s)\left[5 + s + \frac{4}{s}\right] = \frac{8}{s} + 2$$

$$I(s) = \frac{\left(\frac{8}{s} + 2\right)s}{5s + s^2 + 4} = \frac{(2s + 8)}{(s + 4)(s + 1)} = \frac{2}{s + 1}$$

$$i(t) = 2e^{-t}$$

(answer)

$$V_C = \frac{1}{C}\int_0^t i\,dt = \frac{2}{C}\int_0^t e^{-t}\,dt = 8[-e^{-t}]_0^t = 8(1 - e^{-t})$$

If your answer is correct, go to Problem 2-30.
If your answer is not correct, review pp. 15 and 16.

PROBLEM 2-30. DOUBLE ENERGY TRANSIENT

In the circuit below, the capacitor is initially uncharged and the inductance has no stored energy. The resistance is finite but considered negligible for the transient state. Determine the highest instantaneous voltage that will be impressed across the capacitor.

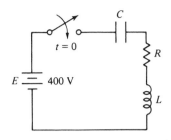

Solution:

Initially, $I_L = 0$ and $V_C = 0$. Summing the voltages around the loop:

$$E = iR + L\frac{di}{dt} + \frac{1}{C}\int_0^t i\,dt$$

converting to Laplace transforms:

$$\frac{E}{s} = \left[R + sL + \frac{1}{sC}\right]I(s)$$

$$I(s) = \frac{E}{s\left[R + sL + \dfrac{1}{sC}\right]} = \frac{E/L}{s^2 + s\dfrac{R}{L} + \dfrac{1}{LC}}$$

$$= \frac{E/L}{\left(s + \dfrac{R}{2L}\right)^2 + \left(\dfrac{1}{LC} - \dfrac{R^2}{4L^2}\right)} \quad = \quad \frac{E/L}{(s+\alpha)^2 + \omega^2}$$

This is the Laplace transform expression for a damped sine wave, where

$$a = -\frac{R}{2L}$$

and

$$\omega^2 = \left(\frac{1}{LC} - \frac{R^2}{4L^2}\right)$$

Converting to the time domain:

$$i(t) = \frac{E}{\omega L}\left[e^{-(R/2L)t}\sin\omega t\right]$$

Assuming R is very small, then the exponential term is negligible and the expression reduces to:

$$i(t) = \frac{E}{\omega L}\sin\omega t \quad \text{and} \quad \omega^2 \approx \frac{1}{LC}$$

The voltage across the capacitor is:

$$V_C = \frac{1}{C} \int_0^t i\, dt = \frac{E}{\omega LC} \int_0^t \sin \omega t\, dt$$

$$= -\frac{E}{\omega^2 LC} [\cos \omega t]_0^t = -E(\cos \omega t - 1)$$

The maximum value of $(\cos \omega t - 1) = (-1 - 1) = -2$

(answer)   Therefore, the maximum value of $V_C = -E(-2) = 2E = 800$ volts.
If your answer is correct, go to Problem 2-31.
If your answer is not correct, review pp. 15 and 16.

PROBLEM 2-31.   TRANSIENT RESPONSE

The switch in the circuit that follows has been in position A for a long time. Determine the equation for $i(t)$ when the switch is moved to position B. Show a plot of $i(t)$ with respect to time.

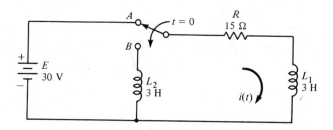

Solution:

Initially, $i(t) = \dfrac{E}{R} = \dfrac{30}{15} = 2$ amps

At $t \geq 0$, excitation function $= E(s) + L_1(I_0) = 0 + 3 \times 2 = 6$

$$Z(s) = sL_1 + sL_2 + R = (L_1 + L_2)\left(s + \frac{R}{L_1 + L_2}\right) = 6(s + 2.5)$$

$$I(s) = \frac{\text{excitation function } (s)}{Z(s)} = \frac{6}{6(s + 2.5)} = \frac{1}{s + 2.5}$$

(answer) $i(t) = e^{-2.5t}$

(answer)

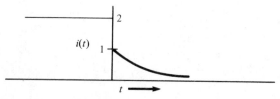

If your answers are correct, go on to Problem 2-32.
If your answers are not correct, review p. 14.

INSULATION

The problem in this section illustrates one aspect of electric fields, a subject of fundamental importance to electrical engineering. Refer to Timbie, Bush, and Hoadley[7] for a good introduction to this subject.

PROBLEM 2-32. INSULATION

A single-conductor cable consists of a copper conductor 0.7 cm in diameter surrounded by a 0.2 cm thick layer of polystyrene insulation and a lead sheath. The insulation has a dielectric constant of 2.6 and a dielectric strength of 300 KV per cm. Determine the maximum safe potential difference that may be applied to the cable.

Solution:

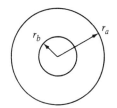

$$V_{ab} = \frac{\Lambda}{2\pi\epsilon} \ln \frac{r_a}{r_b}$$

Λ = linear charge density = $2\pi r_b \epsilon \mathcal{E}$

$\mathcal{E}$ = dielectric constant = 2.6

$\mathcal{E}$ = electric field intensity (= dielectric strength here) = 300 kV/cm

r_a = 0.55 cm

r_b = 0.35 cm

$$V_{ab} = \frac{2\pi r_b \epsilon \mathcal{E}}{2\pi\epsilon} \ln \frac{r_a}{r_b} = r_b \mathcal{E} \ln \frac{r_a}{r_b}$$

$$= 0.35 \text{ cm} \times \frac{300,000 \ V}{\text{cm}} \ln \frac{0.55}{0.35} = 47,560 \text{ volts peak} \qquad \text{(answer)}$$

If your answer is correct, go on to Problem 2-33.

If your answer is not correct, review Timbie[7], section 13-10, or an equivalent text on the field of a linear charge.

WAVEFORMS

Determination of RMS and average values of various types of waveforms is discussed in Chapter 1. The problem in this section illustrates a practical application of this type of analysis.

PROBLEM 2-33. WAVEFORM

A motor draws current in accordance with the periodic waveform below. The cycle repeats every minute and continues for several hours. What must be the continuous current rating of the motor in order to avoid overload?

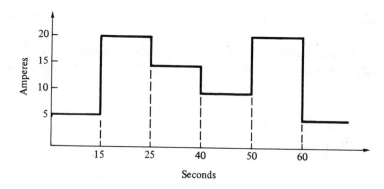

Solution:

The RMS value of the curve is the maximum continuous current capacity of the motor.

$$I_{RMS} = \sqrt{\frac{1}{T} \int_0^t i^2(t)\, dt} =$$

$$= \sqrt{\frac{1}{60}\left[\int_0^{15} 5^2\, dt + \int_{15}^{25} 20^2\, dt + \int_{25}^{40} 15^2\, dt + \int_{40}^{50} 10^2\, dt + \int_{50}^{60} 20^2\, dt \right]}$$

$$= \sqrt{\frac{1}{60}\left[25 \times 15 + 400 \times 10 + 225 \times 15 + 100 \times 10 + 400 \times 10 \right]}$$

(answer) $$= \sqrt{\frac{1}{60}[12{,}750]} = \sqrt{212.5} = 14.58 \text{ amps}$$

If your answer is correct, go on to Chapter 3.
If your answer is not correct, review p. 27.

3 Power

Discussions of and problems related to single-phase KVA and power factors, three-phase power factor and line current, phase sequence, unbalanced load, power factor correction, transmission line, transmission line regulation, and wattmeters.

INTRODUCTION

This chapter covers single phase and polyphase power distribution and loads, power factor correction, transmission line calculations at power frequencies, and wattmeter measurements. Unless otherwise stated, the material in this chapter assumes all waveforms are 60 cycle sine waves. The *vector*, or *phasor*, *diagram* is used to aid in solution of problems.

SINGLE-PHASE POWER

The phase angle is a very important parameter for properly locating different alternating quantities with respect to one another. If the applied voltage is $v = V_{max} \sin \omega t$ and it is known, from the nature and magnitude of the circuit parameters, that the current comes to a corresponding point on its wave before the voltage wave by θ degrees, the current can be expressed as $i = I_{max} \sin (\omega t + \theta)$. This is an example of a positive phase angle due to an RC circuit which produces *leading* current and a leading power factor, where the power factor is:

$$PF = \cos \theta = \frac{\text{watts}}{\text{volt-amperes}} = \frac{G}{|Y|} \quad \text{or} \quad = \frac{R}{|Z|}$$

The phasor diagram for this example is as shown below:

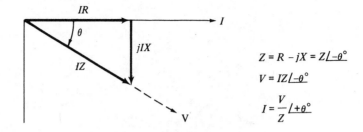

$$Z = R - jX = Z\underline{/-\theta^\circ}$$
$$V = IZ\underline{/-\theta^\circ}$$
$$I = \frac{V}{Z}\underline{/+\theta^\circ}$$

For an *RL* circuit, the phasor diagram is as shown below:

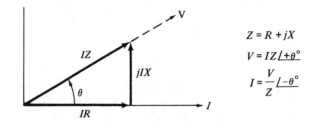

$$Z = R + jX$$
$$V = IZ\underline{/+\theta^\circ}$$
$$I = \frac{V}{Z}\underline{/-\theta^\circ}$$

In this case, the current *lags* the voltage by θ degrees.

Power in ac circuits is either *resistive* (*real*) or *reactive*. Resistive (or average) power, P, is measured in watts, and reactive power, Q, is measured in vars (VAR, for volt-amperes reactive). Since vars are always equal to $VI \sin \theta$ and watts are always equal to $VI \cos \theta$, they can be thought of as being in quadrature to each other. The product VI is called the *apparent* power, S, and is measured in volt-amperes (VA) or kilovolt-amperes (KVA), where

$$VA = \sqrt{\text{watts}^2 + \text{vars}^2}$$

This is represented by the power triangle for an *RL* circuit where current lags voltage.

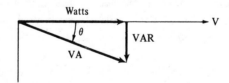

This is a popular diagram for solving power factor correction problems, as discussed later.

For an *RC* circuit, the power phasor diagram would be drawn:

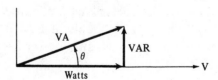

These concepts are illustrated in Problem 3-1 which follows.

PROBLEM 3-1. SINGLE-PHASE KVA AND POWER FACTOR

A machine shop has the following single-phase electrical loads:

- 50 kilowatts incandescent lighting and heating (purely resistive)
- 120 kilowatts at 0.8 power factor lagging
- 25 kilowatts at 0.9 power factor leading

Determine:

 a. the total KVA;
 b. the overall power factor.

Solution:

First, draw a power triangle of each load, observing that they are all given in KW in the problem statement.

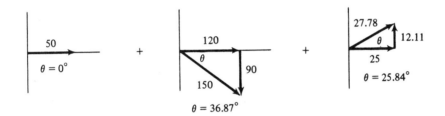

a. KVA = 50 + 120 − j90 + 25 + j12.11

 = 195 − j77.89 = 210$\underline{/-21.77°}$ (answer)

b. PF = cos −21.77° = 0.93 lagging   (answer)

If your answers are correct, go on to the next section.
If your answers are not correct, review pp. 11, 12 and 65.

POLYPHASE POWER

Polyphase voltages are generated the same way as single-phase voltages. A polyphase system is simply several single-phase systems displaced in time phase from one another. In general, the electrical displacement between phases for a balanced, n-phase system is $360/n$ electrical degrees. Three-phase systems are the most common, although for certain special applications, a greater number of phases are used. In general, three-phase equipment is more efficient, uses less material for a given capacity, and costs less than single-phase equipment. Also, for a fixed amount of power to be transmitted a fixed distance at a fixed line loss with a fixed voltage between conductors, three-phase makes more economical use of copper than any other number of phases.

In a balanced three-phase power system, voltages are generated 120° apart. There are two ways of connecting the three phases: one is called the delta (Δ) connection; the other is called the wye (Y) connection. These are shown as follows:

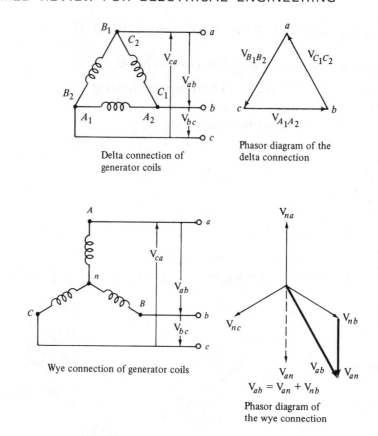

Delta connection of
generator coils

Phasor diagram of the
delta connection

Wye connection of generator coils

Phasor diagram of
the wye connection

The delta phasor diagram shows that the resultant of the voltages in any two phases is equal and opposite to the voltage in the third phase, when all voltages are equal and at 120° to one another, so that there is no net emf around the delta. Thus, $V_{ab} + V_{bc} + V_{ca} = 0$.

In the delta connection the line voltage is equal to the phase voltage. In the wye connection the line voltage is equal to $\sqrt{3}$ times the phase voltage, as shown in the Y-phasor diagram. The Y-connection offers considerable advantage when building high-voltage generators and transformers. Note that in the Y-connection each phase must carry full line current, while in the Δ-connection the phases divide the current between them, each taking $1/\sqrt{3}$ or 0.578 times the line current. Thus, for balanced three-phase, $VA = 3V_{line}I_{line}/\sqrt{3} = \sqrt{3}\ V_{line}I_{line}$.

The figure below shows the voltage waves generated by a three-phase generator. The *phase order* or *sequence* is *abc*, which is the order in which the emf's or phases *a*, *b*, and *c* come to their corresponding maximum values.

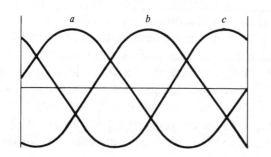

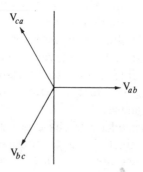

If the rotation of the generator is reversed, the phase sequence would be *acb*. In general, the phase sequence of the voltages applied to a load is fixed by the order in which the three-phase lines are connected. Interchanging any pair of the lines reverses the phase sequence. For three-phase unbalanced loads the effect is to cause a completely different set of values for line currents. Hence, when calculating such systems it is essential that phase sequence be specified.

Unless otherwise stated, the term "phase sequence" refers to *voltage* phase sequence. It should be recognized that, in unbalanced systems, the line currents and phase currents have their own phase sequence which may or may not be the same as the voltage sequence.

The following problems illustrate three-phase calculations for power factor, currents, voltages, power, and phase sequence.

PROBLEM 3-2. THREE-PHASE POWER FACTOR AND LINE CURRENT

Two balanced three-phase loads in parallel with one another are:

1. 50 KVA at 0.707 PF lagging
2. 15 − j20 ohms per phase, delta connection

Determine the power factor and line current of the combined loads if the line voltage is 220 volts.

Solution:

Assuming both loads are delta connected, the load circuit is shown below:

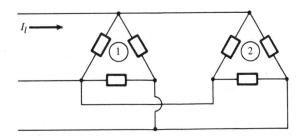

Load 2 is:

$$\frac{V^2}{Z} = \frac{220^2}{25\underline{/-53.13°}} = 1936 \text{ VA @0.6 PF lead per phase}$$

$VA = \sqrt{3} \, I_{LINE} \, V_{LINE}$

Total three-phase load 2 is:

$$3 \times 1936 = 5.808 \text{ KVA @0.6 PF lead}$$

$$\text{Combined load} = 35.36 - j35.36 + 3.485 + j4.646 = 38.83 - j30.71$$

$$= 49.5 \text{ KVA @} - 38.3°$$

$$\text{PF} = \cos - 38.3° = 0.78 \text{ lagging} \qquad \text{(answer)}$$

$$I_{\text{line}} = \frac{VA}{\sqrt{3} V_{\text{line}}} = \frac{49,517}{\sqrt{3} \times 220} = 130 \text{ amps} \qquad \text{(answer)}$$

69

Alternative Solution (impedance method):

$$Z_1 = \frac{3V^2}{VA} = \frac{3 \times 220^2}{50,000\underline{/-45^\circ}} = 2.9\underline{/+45^\circ} = 2.05 + j2.05$$

$$Z_{\|\phi} = \frac{Z_1 Z_2}{Z_1 + Z_2} = \frac{(2.9\underline{/+45^\circ})(25\underline{/-53.13^\circ})}{2.05 + j2.05 + 15 - j20} = \frac{72.6\underline{/-8.13^\circ}}{24.76\underline{/-46.47^\circ}}$$

(answer)
$$= 2.93\underline{/38.3^\circ}; \quad PF = 0.78 \text{ lagging}$$

$$I_\phi = \frac{220}{2.93\underline{/38.34^\circ}} = 75.03\underline{/-38.3^\circ} \text{ amps}$$

(answer)
$$I_{line} = \sqrt{3}\, I_\phi = 130 \text{ amps}$$

If your answers are correct, go on to Problem 3-3.
If your answers are not correct, review pp. 11, 12 and 68.

PROBLEM 3-3. MOTOR INPUT CURRENT

Determine the line current to a 10 horsepower, 60 Hz, three-phase induction motor when it is operating at full load. Motor efficiency is 80%, and power factor is 0.9. Line-to-line input voltage is 480 volts rms.

Solution:

$VA = \sqrt{3}\ I_{line} V_{line}$

$Power = \sqrt{3}\ I_{line} V_{line} \cos\theta$

$$\text{Input power to motor} = \frac{10 \text{ HP}}{0.85} \times \frac{745 \text{ watts}}{\text{HP}} = 8764.7 \text{ watts}$$

$$P = \sqrt{3}\, V_L I_L \cos\theta$$

(answer)
$$I_L = \frac{P}{\sqrt{3}\, V_L \cos\theta} = \frac{8764.7}{\sqrt{3} \times 480 \times 0.9} = 11.7 \text{ amps}$$

If your answer is correct, go on to Problem 3-4.
If your answer is not correct, review p. 68.

PROBLEM 3-4. PHASE SEQUENCE (Lamp Test)

For the unbalanced load circuit shown below, calculate the voltage across the lamps for phase sequences *abc* and *acb*. Line voltage and impedances are:

$$V_{line} = 100 \text{ V}$$

$$Z_{an} = Z_{cn} = 100\underline{/0^\circ}\,\Omega \text{ (resistance)}$$

$$Z_{bn} = 100\underline{/90^\circ} \text{ (inductance)}$$

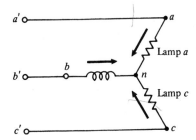

Solution:

Since the system is unbalanced, the neutral point is not at zero volts. For phase sequence *abc*, the voltage phasor diagram is:

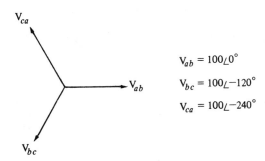

$$V_{ab} = 100 \angle 0°$$
$$V_{bc} = 100 \angle -120°$$
$$V_{ca} = 100 \angle -240°$$

Using Kirchhoff's laws, the following equations may be written:

$$I_{an} + I_{bn} + I_{cn} = 0$$
$$V_{ab} = V_{an} - V_{bn} = I_{an} Z_{an} - I_{bn} Z_{bn}$$
$$V_{bc} = V_{bn} - V_{cn} = I_{bn} Z_{bn} - I_{cn} Z_{cn}$$

Through a series of calculations, it can be shown that:

$$V_{an} = Z_{an} \left[\frac{V_{ab}(Z_{bn} + Z_{cn}) + V_{bc} Z_{bn}}{Z_{an}(Z_{bn} + Z_{cn}) + Z_{cn} Z_{bn}} \right] \qquad \text{(lamp } a\text{)}$$

$$V_{cn} = V_{ca} + V_{an} \qquad \text{(lamp } c\text{)}$$

Substituting in the given values:

$$V_{an} = 100 \left[\frac{100(j100 + 100) + (100\angle -120°)(100\angle 90°)}{100(j100 + 100) + 100(j100)} \right]$$

$$= 100 \left[\frac{10{,}000 + j10{,}000 + 8{,}667 - j5{,}000}{10{,}000 + j10{,}000 + j10{,}000} \right]$$

$$= 100 \left[\frac{18{,}667 + j5000}{10{,}000 + j20{,}000} \right] = 100 \left[\frac{19{,}325\angle 15°}{22{,}361\angle 63.43°} \right] = 86.4\angle -48.43°$$

$$= 86.4 \angle -48.43° \text{ volts across lamp } a \qquad \text{(answer)}$$

$$V_{cn} = 100 \angle -240° + 86.4 \angle -48.43° = 23.15 \angle 71.55° \text{ across lamp } c \qquad \text{(answer)}$$

71

Thus, for sequence *abc*, lamp *a* is brighter than lamp *c*.
For phase sequence *acb*, the voltage phasor diagram becomes:

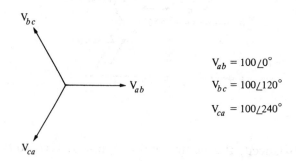

$$V_{ab} = 100\angle 0°$$
$$V_{bc} = 100\angle 120°$$
$$V_{ca} = 100\angle 240°$$

Using the same equations to evaluate the lamp voltages:

$$V_{an} = 100 \left[\frac{100(j100 + 100) + (100\angle 120°)(100\angle 90°)}{22,361\angle 63.43°} \right]$$

$$= 100 \left[\frac{10,000 + j10,000 - 8,667 - j5,000}{22,361\angle 63.43°} \right]$$

(answer)
$$= 100 \left[\frac{5174.64\angle 75.07°}{22,361\angle 63.43°} \right] = 23.14\angle 11.64°$$

(answer)
$$V_{cn} = 100\angle 240° + 23.14\angle 11.64° = 86.4\angle -108.45°$$

Thus, it is seen that by reversing the phase sequence the voltages across the two lamps exchange their magnitudes.

If your answers are correct, go on to Problem 3-5.
If your answers are not correct, review pp. 17 and 68.

PROBLEM 3-5. UNBALANCED LOAD

A three-phase delta-connected load is supplied by a 480 volt power line as shown below:

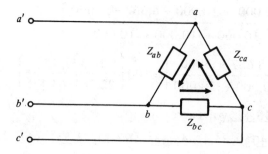

The load impedances are:

$$Z_{ab} = 5 + j5$$

$$Z_{bc} = 3 + j4$$

$$Z_{ca} = 4 - j3$$

Determine the three-phase currents and the current in line a. Show these currents on a phasor diagram.

Solution:

Assume phase sequence abc.

$$I_{ab} = \frac{V_{ab}}{Z_{ab}} = \frac{480\underline{/0^\circ}}{5 + j5} = \frac{480\underline{/0^\circ}}{7.07\underline{/45^\circ}} = 67.88\underline{/-45^\circ} \text{ amps} \qquad \text{(answer)}$$

$$I_{bc} = \frac{V_{bc}}{Z_{bc}} = \frac{480\underline{/-120^\circ}}{3 + j4} = \frac{480\underline{/-120^\circ}}{5\underline{/53.13^\circ}} = 96\underline{/-173.13^\circ} \text{ amps} \qquad \text{(answer)}$$

$$I_{ca} = \frac{V_{ca}}{Z_{ca}} = \frac{480\underline{/+120^\circ}}{4 - j3} = \frac{480\underline{/120^\circ}}{5\underline{/-36.87^\circ}} = 96\underline{/156.87^\circ} \text{ amps} \qquad \text{(answer)}$$

$$I_{a'a} = I_{ab} - I_{ca} = 67.88\underline{/-45^\circ} - 96\underline{/156.87^\circ} = 160.99\underline{/-32.17^\circ} \text{ amps} \qquad \text{(answer)}$$

The phasor diagram is shown below:

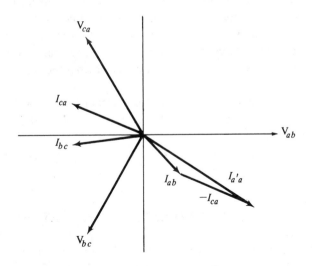

If your answers are correct, go on the the next section.
If your answers are not correct, review pp. 5, 11, 12 and 68.

POWER FACTOR CORRECTION

Most industrial loads contain many devices such as induction motors, which have a lagging, or inductive, power factor. This power factor ranges from 0.9 to 0.7 in most cases. The effect of a low power factor is to require more current for the same power. Another undesirable effect is that it also causes transmission lines to have larger regulation. To correct this situation, capacitance is added in parallel with the load as shown below:

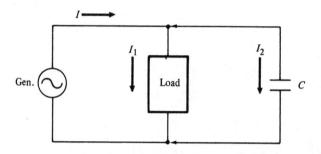

A phasor diagram of power factor correction in which θ_1 is improved to θ is shown below:

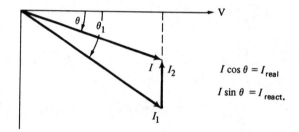

$$I \cos \theta = I_{real}$$
$$I \sin \theta = I_{react.}$$

Power factor correction presents certain problems. For example, if the load changes, the resultant power factor will probably change, so it may be necessary to regulate the size of I_2, perhaps with automatic equipment, to keep the power factor at a steady value (not necessarily unity). This assures that the voltage is constant at the load. Sometimes each motor in a plant is equipped with its own capacitor, which is connected to the line only when the motor is connected.

Two types of devices are used in power factor correction: a capacitor or a synchronous motor. When a capacitor is used, it must be able to withstand the peak voltage ($\sqrt{2}$ times the RMS voltage) twice each cycle.

The following problem illustrates a power factor correction calculation.

PROBLEM 3-6. POWER FACTOR CORRECTION (CAPACITOR)

A three-phase feeder line supplies a lighting load of 500 KW at unit power factor and a motor load of 1500 KVA at 0.7 power factor. Calculate the KVAR of shunt capacitance required to correct the power factor to 0.9.

Solution:

Draw the uncorrected power factor phasor diagram:

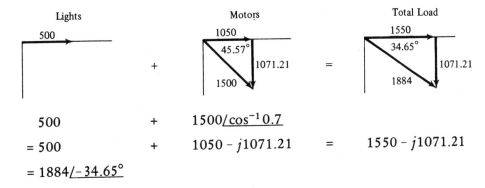

$$500 \quad + \quad 1500\underline{/\cos^{-1}0.7}$$

$$= 500 \quad + \quad 1050 - j1071.21 \quad = \quad 1550 - j1071.21$$

$$= 1884\underline{/-34.65°}$$

We would like the corrected load to be:

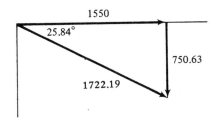

The difference between corrected and uncorrected KVAR's is:

$$1071.21 - 750.63 = 320.6 \text{ KVAR}$$

This is the required KVAR amount of shunt capacitance.   (answer)

If your answer is correct, go on to Problem 3-7.
If your answers are not correct, review pp. 66, 67, and 74.

TRANSMISSION LINE CALCULATIONS

A transmission line consists of the equivalent of two or more electrical conductors for the purpose of transmitting electrical energy. For single-phase transmission the line may consist of a single conductor with a ground return, or of two ordinary wires. For three-phase transmission, three wires are generally used, although in some installations a neutral wire or its equivalent is employed. The figure below shows a *lumped-parameter* transmission line which is a good approximation to a real transmission line. V_s and I_s denote the sending-end voltage and current, and V_r and I_r represent receiver-end voltage and current.

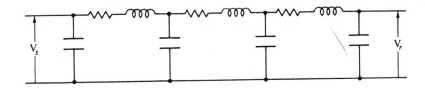

Two popular arrangements for calculation are the T-line and the Π-line. The T-line representation of a transmission line is shown below:

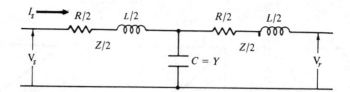

When all the shunted capacitance C of the line is concentrated in one capacitor and one-half of the total series impedance Z is placed in each arm as indicated, the circuit is known as the nominal T-line. Calling Y the admittance due to shunted capacitance, formulas for V_s and I_s can be derived to give the following results:

$$V_s = AV_r + BI_r$$

$$I_s = CV_r + DI_r$$

where

$$A = 1 + \frac{YZ}{2}$$

$$B = Z + \frac{YZ^2}{4}$$

$$C = Y$$

$$D = A$$

The Π-line representation of a transmission line is as follows:

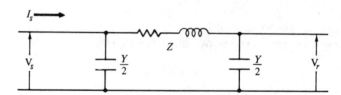

In this case, one-half of the total line capacitance is concentrated at each end of the line, and all the series resistance and reactance are concentrated at the center. Formulas for V_s and I_s in this case are:

$$V_s = AV_r + BI_r$$

$$I_s = CV_r + DI_r$$

where

$$A = 1 + \frac{ZY}{2}$$

$$B = Z$$

$$C = Y\left(1 + \frac{ZY}{4}\right)$$

$$D = A$$

PROBLEM 3-7. POWER FACTOR CORRECTION (SYNCHRONOUS MOTOR)

An overexcited synchronous motor is to be used to improve the power factor of a 2000 KW load having a lagging power factor of 0.83. The power factor of the motor will be adjusted to 0.8. If the resultant power factor is to be 0.94 lagging, what must be the KVA rating of the synchronous motor?

Solution:

Drawing the power triangles:

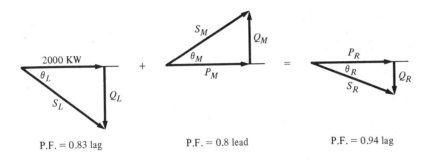

The problem is to calculate S_M, the motor apparent power. Write two equations, one for reals and one for imaginaries:

(1) $2000 + P_M = P_R$
(2) $Q_L - Q_M = Q_R$

Calculate the phase angle, S_L and Q_L:

$$\theta_L = \text{arc cos } 0.83 = 33.9°$$

$$S_L = \frac{2000}{0.83} = 2409.64 \text{ KVA}$$

$$Q_L = 2409.64 \sin 33.9° = 1344 \text{ KVAR}$$

Calculating the motor and resultant phase angles:

$\theta_M = \text{arc cos } 0.8 = 36.87°$
$\theta_R = \text{arc cos } 0.94 = 19.95°$

$\tan \theta_R = \dfrac{Q_R}{P_R}$, $Q_R = P_R \tan 19.95° = 0.36 P_R$

$\tan \theta_M = \dfrac{Q_M}{P_M}$, $Q_M = P_M \tan 36.87° = 0.75 P_M$

Substituting into equations (1) and (2):

(1) $2000 + P_M = P_R$
(2) $1344 - 0.75 P_M = 0.36 P_R$

77

Multiplying (1) by 0.75 and adding to (2):

$$1500 + 0.75P_M = 0.75P_R$$
$$\underline{1344 - 0.75P_M = 0.36P_R}$$
$$2844 \qquad\qquad = 1.11P_R$$

$$P_R = \frac{2844}{1.11} = 2562 \text{ KW}$$

From (1):

$$P_M = P_R - 2000 = 2562 - 2000 = 562 \text{ KW}$$

Solving for Q_M and Q_R:

$$Q_M = 0.75P_M = 422 \text{ KVAR}$$
$$Q_R = 0.36P_R = 922 \text{ KVAR}$$

Solving for S_R and S_M:

$$S_R = \frac{P_R}{\cos\theta_R} = \frac{2562}{0.94} = 2726 \text{ KVA}$$

(answer)

$$S_M = \frac{P_M}{\cos\theta_M} = \frac{562}{0.8} = 703 \text{ KVA}$$

If your answer is correct, go on to Problem 3-8.
If your answer is not correct, review pp. 66 and 67.

PROBLEM 3-8. POWER FACTOR (ADJUSTING EXISTING LOAD)

A small industrial company is being charged a premium rate for electricity because of poor power factor. Its electrical load of 500 KW has a lagging power factor of 0.75. However, the company is operating three 100 HP synchronous motors at unity power factor for maximum efficiency.

If the fields were to be overexcited, the motors could be made to operate at a leading power factor (at some small loss in efficiency) which would improve overall power factor. If the power factor could be improved to greater than 0.85, the premium electricity rate would be eliminated. Normally, the motors operate at 80% of capacity. However, the fields can be overexcited so that each motor is drawing rated current while still delivering 80% of rated capacity.

Determine the new power factor if the motors are operated in this manner.

Solution:

Assume motors' efficiency is 85%. Maximum motor load is:

$$3 \times \frac{100 \text{ HP}}{0.85} \times \frac{0.746 \text{ KW}}{\text{HP}} = 263.3 \text{ KVA}$$

The motors are presently drawing 263.3 × 0.8 = 210.64 KW. Drawing the power vectors:

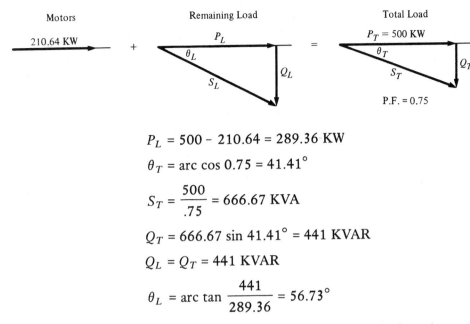

$$P_L = 500 - 210.64 = 289.36 \text{ KW}$$

$$\theta_T = \text{arc cos } 0.75 = 41.41°$$

$$S_T = \frac{500}{.75} = 666.67 \text{ KVA}$$

$$Q_T = 666.67 \sin 41.41° = 441 \text{ KVAR}$$

$$Q_L = Q_T = 441 \text{ KVAR}$$

$$\theta_L = \text{arc tan } \frac{441}{289.36} = 56.73°$$

Calculating the power triangle parameters of the synchronous motors when they are made to draw full load line current and output 80% of rated capacity:

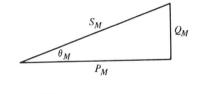

$$S_M = 263.3 \text{ KVA}$$

$$P_M = 210.64 \text{ KW}$$

$$\theta_M = \text{arc cos } \frac{210.64}{263.3} = 36.87°$$

$$Q_M = 263.3 \sin 36.87 = 157.98 \text{ KVAR}$$

$$Q_T = Q_L - Q_M = 441 - 157.98 = 283.02$$

$$S_T = P_T + jQ_T = 500 + j283.02 = 574.54 \underline{/29.51°}$$

$$\text{P.F.}_T = \cos 29.51° = 0.87 \qquad \text{(answer)}$$

This is above the 0.85 minimum, so normal electric rates will apply.

If your answer is correct, go on to the next section.
If your answer is not correct, review pp. 66 and 67.

PROBLEM 3-9. TRANSMISSION LINE

A 60 Hz, 3-phase line 200 miles long has a shunt capacitance to neutral per mile of 150×10^{-4} μf, an inductive reactance of 0.75 ohm per wire per mile, and a resistance of 0.5 ohm per wire per mile. The receiver voltage is 120 kilovolts between lines. Find the sending end voltage and current for an 0.8 power factor lagging load requiring 70 amps per line at the receiver.

Solution:

Arbitrarily select the Π-line configuration.

$$V_s = AV_r + BI_r \qquad A = 1 + \frac{ZY}{2}$$

$$I_s = CV_r + DI_r \qquad B = Z$$

$$C = Y\left(1 + \frac{ZY}{4}\right)$$

$$D = A$$

$$Z = (R + j\omega L)d = (R + jX_L)d$$
$$Y = j\omega Cd = j(1/X_C)d$$

Given (per phase):

$R = 0.5\ \Omega/\text{mile}$
$C = 150 \times 10^{-4}\ \mu f/\text{mile}$
$X_L = 0.75\ \Omega/\text{mile}$
$d = 200$ miles
$V_r = 120$ KV line-to-line
PF = 0.8 lag
$I_r = 70$ amps/line
$f = 60$ Hz

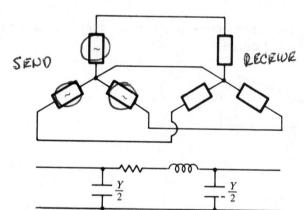

$$Z = (0.5 + j0.75)200 = 0.9\underline{/56.31°} \times 200 = 180.28\underline{/56.31°}$$

$$Y = j2\pi \times 60 \times 150 \times 10^{-10} \times 200 = 1.131 \times 10^{-3}\underline{/90°}$$

$$ZY = 180.28\underline{/56.31°} \times 1.131 \times 10^{-3}\underline{/90°} = 0.20389\underline{/146.31°} = -0.1696 + j0.1131$$

$$A = 1 - 0.0848 + j0.05655 = 0.9152 + j0.05655 = 0.9169\underline{/3.536°}$$

$$B = 180.28\underline{/56.31°}$$

$$C = (1.131 \times 10^{-3}\underline{/90°})(1 - 0.0424 + j0.0283) = (1.131 \times 10^{-3}\underline{/90°})(0.9576 + j0.0283)$$
$$= 0.0010835\underline{/91.69°}$$

$$V_{s\phi} = 0.9169\underline{/3.536°} \times \frac{120 \times 10^3}{\sqrt{3}} + 180.28\underline{/56.31°} \times 70\underline{/-36.9°}$$

$$= 6.353 \times 10^4\underline{/3.54°} + 1.262 \times 10^4\underline{/19.44°}$$

$$= 75{,}746\underline{/6.16°} \text{ volts line-to-neutral}$$

Converting to line voltage:

$$V_s = \sqrt{3}\ 75,746\underline{/6.16°} = 131,196\underline{/6.16°} \text{ volts line-to-line} \qquad \text{(answer)}$$

$$I_s = 0.0010835\underline{/91.69°} \times \frac{120 \times 10^3}{\sqrt{3}} + 0.9169\underline{/3.536°} \times 70\underline{/-36.9°}$$

$$= 75.067\underline{/91.69°} + 64.183\underline{/-33.334°}$$

$$= 64.994\underline{/37.72°} \text{ amps} \qquad \text{(answer)}$$

If the problem had been worked using the T-line configuration, the results would have been:

$$V_s = 130,142\underline{/6.32°} \text{ volts} \quad \text{and} \quad I_s = 68.79\underline{/38.8°} \text{ amps.}$$

The difference is due to the fact that neither method is exact.

Alternate Solution:

The following solution is applicable to lines of 150 miles or less. Even though this line is 200 miles long, it may be interesting to compare the results.

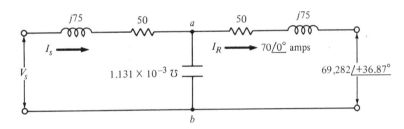

I_R is referenced to $0°$.

$$\begin{aligned} V_{ab} &= V_R + I_R Z = 69,282\underline{/+36.87°} + 70\,(50 + j75) \\ &= 55,426 + j41,569 + 3500 + j5250 \\ &= 58,926 + j46,819 = 75,261.5\underline{/38.47°} \end{aligned}$$

$$\begin{aligned} I_{ab} &= V_{ab}Y_{ab} = (75,261.5\underline{/38.47°})(1.131 \times 10^{-3}\underline{/90°}) \\ &= 85.12\underline{/128.47°} = -52.95 + j66.64 \end{aligned}$$

$$\begin{aligned} I_s &= I_R + I_{ab} = 70 - 52.95 + j66.64 = 17.05 + j66.64 \\ &= 68.79\underline{/75.65°} \end{aligned}$$

$$\begin{aligned} V_{s\phi} &= V_{ab} + I_s Z = 75,261.5\underline{/38.47} + (68.79\underline{/75.65})(50 + j75) \\ &= 58,924.8 + j46,820.5 - 4145.8 - j4610.9 \\ &= 54,779 + j51,431.4 = 75,139.4\underline{/43.19°} \end{aligned}$$

$$V_s = \sqrt{3}\ V_{s\phi} = 130,145 \text{ volts} \qquad \text{(answer)}$$

This method yields nearly the same results as the T-line configuration.

If your answers are correct, go on to Problem 3-10.
If your answers are not correct, review pp. 75 and 76.

PROBLEM 3-10. TRANSMISSION LINE REGULATION

A balanced three-phase load is supplied by a 3-wire transmission line having a series impedance of $3 + j4$ ohms in each line. The line voltage is 10 KV at the load. Power dissipated by the load is 3000 KW at 75% lagging power factor. Calculate the voltage regulation of the line.

Solution:

Regulation is defined as the difference between the full-load and no-load voltage at the load terminals and is expressed as a percentage of the full-load voltage as follows:

$$\% \text{ Regulation} = \frac{|V_{nl}| - |V_{fl}|}{|V_{fl}|} \times 100 = \frac{|V_s| - |V_r|}{|V_r|} \times 100$$

This formula is commonly used in calculating transformer regulation, as will be discussed in Chapter 4. In this case, no-load voltage is the same as the sending end voltage, and full-load voltage is the sending end voltage minus the drop in the line. The circuit may be represented by the figure below:

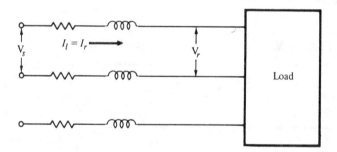

$$P = \sqrt{3} V_l I_l \cos \theta, \quad V_r = 10 \text{ KV}, \quad P = 3000 \text{ KW}, \quad \cos \theta = 0.75 \text{ lag}$$

$$I_l = \frac{P}{\sqrt{3} V_l \cos \theta} = \frac{3000 \times 10^3}{\sqrt{3} \times 10^4 \times 0.75} = 231\underline{/-41.4°} \text{ amps}$$

$$V_{s\phi} = \frac{V_r}{\sqrt{3}} + I_r Z_{\text{line}} = \frac{10,000}{\sqrt{3}} + 231\,(0.75 - j0.66)\,(3 + j4)$$

$$= 5773.5 + 1154.7\underline{/11.73°} = 6908\underline{/1.95°} \text{ volts line-to-neutral}$$

Converting to line voltage:

$$V_s = \sqrt{3} \, V_{s\phi} = \sqrt{3} \times 6908\underline{/1.95°} = 11,965\underline{/1.95°} \text{ volts line-to-line}$$

$$\% \text{ Regulation} = \frac{11,965 - 10,000}{10,000} \times 100 = \frac{1,965}{10,000} \times 100 = 19.65\% \quad \text{(answer)}$$

This is an example of very poor regulation.

If your answer is correct, go on to the next section.
If your answer is not correct, review pp. 99 and 111.

WATTMETER MEASUREMENTS

A wattmeter gives a reading proportional to the product of the current through its current coil, the voltage across its potential coil, and the cosine of the angle between the voltage and current. Since the total power in a three-phase circuit is the sum of the powers of the separate phases, the total power could be measured by placing a wattmeter in each phase. Since this method is undesirable unless the individual phase powers are required, another method making use of only two wattmeters is generally employed in making three-phase power measurements, as shown below:

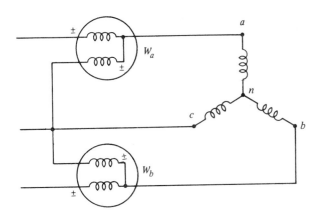

The power measured by each wattmeter is given by the formulas:

$$W_a = V_{ac}I_{an} \cos (\theta - 30°)$$

$$W_b = V_{bc}I_{bn} \cos (\theta + 30°)$$

$$W_a + W_b = VI \cos (\theta - 30°) + VI \cos (\theta + 30°) = \sqrt{3}\,V_l I_l \cos \theta$$

Hence, $W_a + W_b$ correctly measures the power in a balanced three-phase system of any power factor. Furthermore, the algebraic sum of the readings of two wattmeters will give the correct value for power under any conditions of unbalance, waveform, or power factor.

The direction of rotation of polyphase induction motors is dependent upon the phase sequence of the applied voltages. Similarly, two wattmeters in the two-wattmeter method of measuring three-phase power interchange their readings when subjected to a reversal of phase sequence, even though the system is balanced. In general, $n - 1$ wattmeters can be used to measure n-phase power. Note that the return lines of all voltmeter windings must be connected to the same point.

PROBLEM 3-11. WATTMETER

Two single-phase wattmeters are connected in a balanced three-phase circuit, as shown below. The three load impedances are $Z = 4 + j3$, and the line voltage is 440 volts. Calculate the wattmeter readings for a phase sequence of abc.

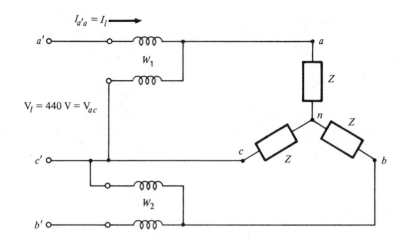

Solution:

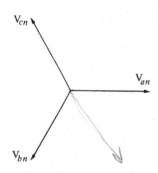

$$V_{an} = \frac{440}{\sqrt{3}} = 254 \text{ volts} = V_\phi$$

$$I_l = I_\phi = \frac{V_\phi}{Z} = \frac{254}{5\underline{/36.87°}} = 50.81 \text{ amps}$$

$$P_\phi = I^2 R = 50.81^2 \times 4 = 10{,}325 \text{ watts}$$

$$P_{tot} = 3P_\phi = 3 \times 10{,}326.6 = 30{,}976 \text{ watts}$$

or

$$P_{tot} = \sqrt{3}\, V_l I_l \cos\theta = \sqrt{3} \times 440 \times 50.81 \times \cos 36.87° = 30{,}976 \text{ watts}$$

This represents the algebraic sum of the two wattmeter readings. Now, calculating the wattmeter readings:

$$W_1 = W_{ac-a'a} = V_{ac} I_{a'a} \cos\theta \Big|_{I_{a'a}}^{V_{ac}}$$

The notation $\cos\theta \Big|_{I_{a'a}}^{V_{ac}}$ represents the angle between line voltage V_{ac} and line current $I_{a'a}$.

$$V_{ac} = V_{an} + V_{nc} = 254\underline{/0°} + 254\underline{/-60°} = 440\underline{/-30°} \text{ volts}$$

$$I_{a'a} = 50.81\underline{/-36.87°} \text{ amps}$$

(answer)
$$W_1 = 440 \times 50.81 \cos 6.87° = 22{,}194.5 \text{ watts}$$

$$W_2 = W_{bc-b'b} = V_{bc} I_{b'b} \cos\theta \Big|_{I_{b'b}}^{V_{bc}}$$

$$V_{bc} = V_{bn} + V_{nc} = 254\underline{/-120°} + 254\underline{/-60°} = 440\underline{/-90°} \text{ volts}$$

$$I_{b'b} = \frac{V_{bn}}{Z_{bn}} = \frac{254\underline{/-120°}}{5\underline{/36.87°}} = 50.81\underline{/-156.87°}$$

(answer)
$$W_2 = 440 \times 50.81 \cos 66.87° = 8781.5 \text{ watts}$$

Check:

$$P_{tot} = W_1 + W_2 = 30{,}976 \text{ watts}$$

This agrees with the initial calculation.

84

An alternative solution is to use the equation given in the preceding text:

$$W_1 = V_{ac}I_{an} \cos(\theta - 30°)$$

$$= 440 \times 50.81 \cos(36.87 - 30°) = 22,194.5 \text{ watts}$$

$$W_2 = V_{bc}I_{bn} \cos(36.87 + 30°) = 8781.5 \text{ watts}$$

These are identical to the previous calculations; the only difference is the manner in which the cosines of the angles are obtained.

If your answers are correct, go on to Problem 3-12.
If your answers are not correct, review p. 83.

PROBLEM 3-12. WATTMETER (UNBALANCED LOAD)

Determine the reading of the wattmeter shown connected to the unbalanced three-phase load that follows. Line voltage is 208 V_{rms}, and the phase sequence is c-b-a.

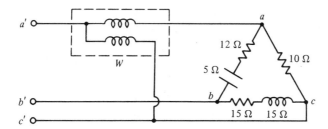

Solution:

The phasor diagram of the three line voltages referenced to V_{ac} is:

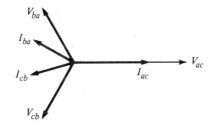

$$P = V_{ac}I_{a'a} \cos \theta \left| \begin{matrix} V_{ac} \\ I_{a'a} \end{matrix} \right.$$

Evaluating all the currents for future use in checking the solution:

$$I_{ab} = \frac{V_{ab}}{Z_{ab}} = \frac{208\underline{/-60°}}{13\underline{/-22.62°}} = 16\underline{/-37.38°} = 12.71 - j9.71 \text{ amps}$$

$$I_{bc} = \frac{V_{bc}}{Z_{bc}} = \frac{208\underline{/+60°}}{21.21\underline{/+45°}} = 9.81\underline{/15°} = 9.48 + j2.54 \text{ amps}$$

$$I_{ca} = \frac{V_{ca}}{Z_{ca}} = \frac{208\underline{/-180°}}{10} = 208\underline{/-180°} = -20.8 \text{ amps}$$

$$I_{a'a} = I_{ab} + I_{ac} = 12.71 - j9.71 + 20.8 = 33.51 - j9.71 = 34.89 \underline{/-16.16°}$$

$$I_{b'b} = I_{ba} + I_{bc} = 12.71 + j9.71 + 9.48 + j2.54 = -3.23 + j12.25 = 12.67 \underline{/104.77°}$$

$$I_{c'c} = I_{cb} + I_{ca} = -9.48 - j2.54 - 20.8 = -30.28 - j2.54 = 30.39 \underline{/-175.21°}$$

Calculating the individual phase power dissipations and summing them:

$$P_{ab} = V_{ab}I_{ab} \cos \theta = (208\underline{/-60°})(16\underline{/-37.38°}) \cos 22.62 = 3072 \text{ watts}$$

$$P_{bc} = V_{bc}I_{bc} \cos \theta = (208\underline{/+60°})(9.81\underline{/15°}) \cos 45° = 1443 \text{ watts}$$

$$P_{ca} = V_{ca}I_{ca} \cos \theta = (208\underline{/-180°})(20.8\underline{/-180°}) \cos 0° = 4326 \text{ watts}$$

$$P_T = 8844 \text{ watts}$$

Now, returning to the calculation of the wattmeter reading:

$$\text{ℓ, } \quad P = V_{ac}I_{a'a} \cos \theta \Big|_{I_{a'a}}^{V_{ac}} = (208°\underline{/0°})(34.89\underline{/16.16°}) \cos 16.16°$$

(answer)
$$= 6970 \text{ watts}$$

Check:

Calculate the reading of a second wattmeter connected to phase b with its voltage winding return connected to the same line as the first wattmeter, namely line phase c.

$$P = V_{bc}I_{b'b} \cos \theta \Big|_{I_{b'b}}^{V_{ab}} = (208\underline{/+60°})(12.67\underline{/+104.71°}) \cos 44.71°$$

$$= 1873 \text{ watts}$$

Adding the two wattmeter readings yields the same result as the three-phase calculation of 8844 watts.

If your answer is correct, go on to the next section.
If your answer is not correct, review pp. 66 and 67.

SHORT CIRCUIT CALCULATIONS

Power systems are susceptible to three kinds of short circuits:
Three-phase—all three lines of a three-phase system electrically connected;
Line-to-line—two lines electrically connected;
Line-to-ground—a single wire electrically connected to ground.

Short circuits are called *faults*. By use of a relay-operated circuit breaker, a distribution system may be protected from a faulty section. Circuit breakers are designed to trip under the lowest current fault condition, yet handle worst case (greatest) fault currents.

A distribution network consists of many lines which may be connected by transformers and which generally operate at different voltages. To establish a simple network for purposes of calculation, the impedances of all lines and transformers are expressed in ohms referred to a common voltage base, or in percentage (or per unit) referred to a common KVA base. The latter method is preferred by most power engineers.

Percentage reactance is defined as the percentage of rated voltage which is consumed in a reactance drop when rated current flows. Expressed algebraically:

$$\% \text{ reactance} = \frac{I_{\text{rated}} \times \text{ohms}}{V_{\text{rated}}} \times 100$$

Percent resistance is similarly defined. Percentage values are manipulated like ohmic values.

By way of example, consider the following per-phase equivalent circuit of a three-phase system:

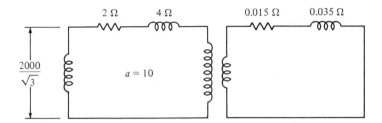

The corresponding one-line diagram in which a fault is indicated on the secondary is as follows:

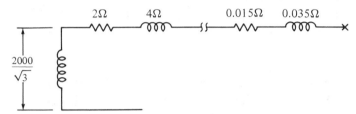

Referring the secondary to the primary, the following equivalent current results, assuming a transformer impedance per phase referred to the primary of $1 + j2$ ohms:

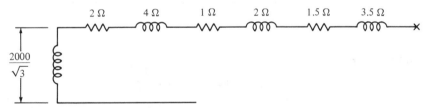

Converting from ohmic values to percentage values (Assume a 10,000 KVA primary base and a 100 KVA secondary base.):

$$\text{base current } I = \frac{10,000,000}{\sqrt{3} \times 2,000} = 2887 \text{ amps}$$

$$\% \text{ } IX \text{ drop due to base current} = 100 \times \frac{2887 \times 4}{2000/\sqrt{3}} = 1000$$

$$\% \text{ } IR \text{ drop due to base current} = 100 \times \frac{2887 \times 2}{2000/\sqrt{3}} = 500$$

Transformer impedance on 10,000 KVA base is:

$$\% \text{ } IR \text{ drop} = 100 \times \frac{2887 \times 1}{2000/\sqrt{3}} = 250$$

$$\% \text{ } IX \text{ drop} = 100 \times \frac{2887 \times 2}{2000/\sqrt{3}} = 500$$

Secondary line impedance based on 100 KVA is:

$$\text{secondary voltage} = \frac{2000/\sqrt{3}}{10} = 200/\sqrt{3} \text{ volts}$$

$$\text{base current } I = \frac{100,000}{\sqrt{3} \times 200} = 288.7 \text{ amps}$$

$$\% \, IX \text{ drop} = 100 \times \frac{288.7 \times 0.035}{200/\sqrt{3}} = 8.75$$

$$\% \, IR \text{ drop} = 100 \times \frac{288.7 \times 0.015}{200/\sqrt{3}} = 3.75$$

A one-line diagram with parameters expressed on a percentage basis is as follows:

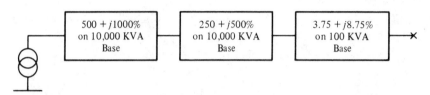

Now choose a common KVA base to which all constants may be referred. For discussion, arbitrarily select 1000 KVA. Since impedance varies directly with the KVA base, the following schematic results:

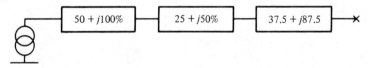

The combined impedance to the fault is:

$$Z = 112.5 + j237.5\% = 263 \underline{/65°} \, \%$$

Thus, **263%** of rated voltage is required to cause 1000 KVA to be delivered by the generator. Since only rated voltage, or 100% voltage, is available, the total short circuit KVA must be $\frac{100}{263} \times 1000 = 380.5$ KVA. Fault current at the actual voltage of the faulty line is calculated as follows:

$$I_{\text{fault}} = \frac{380.5 \times 1000}{200\sqrt{3}} = 1098.4 \text{ amps}$$

Continue on to Chapter 4.

4 Machinery

Discussions of and problems related to series-wound dc motors, shunt-wound dc motors, series-shunt-wound dc motors, shunt motor torque, induction motor speed, induction motor efficiency, induction motor losses, motor starting—line voltage drop, transformer efficiency and regulation, regulation improvement, autotransformers, and magnetic devices.

INTRODUCTION

Electrical machinery may be divided into two categories: dc and ac. Dc machinery consists, broadly, of motors and generators, whereas ac machinery consists of motors, generators, and transformers.

Most PE Exam problems deal with motors and transformers. Generators are almost totally ignored since their characteristics, in principle, are nearly identical to those of motors. Many varieties of motors are available to fit the myriad applications.

Table 4-1 categorizes basic electric motors by type. Detailed characteristics are summarized in handbooks available from many motor manufacturers. Transformers also may be classified by phase (single phase or polyphase) and type (two-winding or autotransformer). The problems in this chapter illustrate the features and parameters of rotating machinery and transformers.

DC MACHINES

Dc motors and generators consist of an *armature*, a *field*, a *commutator*, and *poles*. The typical dc generator uses stationary electromagnets for producing the fields. Conductors for the generation of emf are carried on a rotating element called the armature.

TABLE 4-1.   Electric Motor Classification.

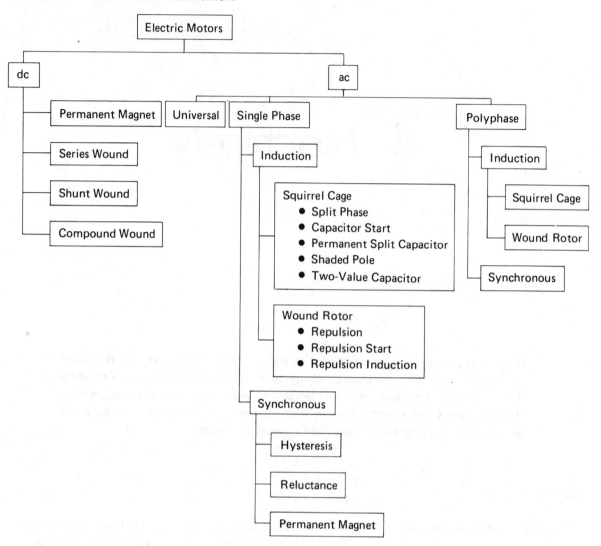

The instantaneous emf induced in a conductor of length l moving with a velocity v within and perpendicular to a magnetic field of density B is:

$$e = Blv$$

Also, the force on a conductor of length l when carrying a current i within and perpendicular to a magnetic field of density B is:

$$f = Bil \begin{cases} \text{newtons} = \dfrac{\text{webers}}{m^2} \times \text{amps} \times \text{m} & \text{(RMKS)} \\[2mm] \text{dynes} = 10\ \text{gauss} \times \text{amps} \times \text{cm} & \text{(CGS)} \\[2mm] \text{pounds} = \dfrac{\text{lines}}{m^2} \times \text{amps} \times \dfrac{\text{in.}}{0.885 \times 10^{-7}} & \text{(ENGLISH)} \end{cases}$$

90

Dividing the first equation by the second yields:

$$ei = fv$$

or

$$\text{electrical power} = \text{mechanical power}$$

Because of the reversible property of the energy conversion, it is possible for the same machine to act as either a generator or a motor.

A conductor of N turns moving across a magnetic field has an emf generated in it equal to:

$$E = BlvN \times 10^{-8} \text{ volts}$$

where

B is in lines/in.^{2}
l is in in.
v is in in./sec
N is number of turns

If the flux linking a *coil* of wire changes, an emf will be induced in that coil in accordance with the equation:

$$E = -N\frac{d\phi}{dt} \times 10^{-8}$$

The coil and field poles representing motor action are identical with those representing generator action. This fact gives rise to the term *back* or *counter emf* for the generated emf when a machine is acting as a motor. A change from motor to generator action for the same direction of rotation and polarity of the field is due only to a reversal of armature current. In the generator, the current flows in the direction of the induced emf, while in the motor it is opposite.

When considering the difference between electrical degrees and mechanical degrees, the number of poles must be taken into consideration. In ac machinery this concept is important because frequency enters in. A conductor must pass two poles to produce one cycle. Hence, the number of electrical degrees per revolution of the armature shaft is:

$$\frac{\text{electrical degrees}}{\text{revolution}} = \frac{\text{no. of poles}}{2} \times 360$$

The terminal voltage of a generator tends to drop when the machine is loaded. The term *voltage regulation* is applied specifically to this situation and is defined by the following equation:

$$\% \text{ voltage regulation} = \frac{(\text{no-load voltage}) - (\text{full-load voltage})}{\text{full-load voltage}} \times 100$$

The terminal voltage of a generator under load differs from the internal generated voltage by the amount of potential drop in the armature series circuit. The resistance drop is due to the total resistance encountered by the load current as it flows between the – and + terminals of the machine. This total resistance includes that of the brushes, brush contact, armature, series field, interpole field, and compensating windings. Thus, the terminal potential V_t is always equal to the generated voltage E minus the potential drop $I_a R_a$ in the armature series circuit as expressed by the equation:

$$V_t = E - I_a R_a$$

where the machine is operating as a *generator*. For *motor action* the sign of the *IR* drop changes (since the direction of I changes) and E becomes the back emf.

A motor develops torque because of the emf exerted on a conductor when it is carrying current in a magnetic field. The torque can be expressed by the equations:

$$T = \frac{7.045}{S} EI_a = \frac{33,000 \text{ HP}}{2\pi S}$$

where

HP   is gross mechanical hp of motor = $EI_a / 746$
S is speed of motor in rpm
T is in pound-feet

Speed regulation of a motor when operating at constant terminal potential is defined by the expression:

$$100 \times \frac{\text{(no-load speed)} - \text{(full-load speed)}}{\text{full-load speed}} \%$$

Efficiency is defined as output divided by input. For a generator,

$$\eta_G = \frac{\text{output}}{\text{output} + \text{losses}} = 1 - \frac{\text{losses}}{\text{output} + \text{losses}}$$

For a motor,

$$\eta_M = \frac{\text{input} - \text{losses}}{\text{input}} = 1 - \frac{\text{losses}}{\text{input}}$$

Losses are grouped as follows:

    I. Electrical losses
       A. Ohmic losses
          1. $I^2 R_f$ loss in shunt field winding
          2. $I^2 R_a$ loss in armature winding
          3. $I^2 R_s$ loss in series field winding
          4. $I^2 R_{\text{rheo}}$ loss in rheostat
       B. Brush contact loss ($V_{\text{brush}} I_a$)
   II. Rotational losses
       A. Mechanical
          1. Friction and windage
          2. Brush friction
       B. Core loss (iron loss, hysteresis, and eddy current)
       C. Ventilating loss
       D. Stray-load loss

The all-day efficiency of a machine is defined as

$$\frac{\text{output}}{\text{output + constant losses + variable losses}}$$

Efficiency is high under full load and low at light loads, and this loading usually varies during a 24-hour period.

PROBLEM 4-1. SERIES-WOUND DC MOTOR

A series-wound dc motor is supplied from a source of E volts having an internal resistance of R_g. The motor generates a back emf E_m (proportional to its speed) in series with its armature and field windings represented by resistor R_m as shown in the following circuit:

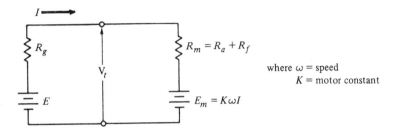

where ω = speed
K = motor constant

Determine the maximum average power that can be converted to mechanical form and the motor speed at which this occurs.

Solution:

Voltage at the motor terminals is given by the equation:

$$V_t = IR_m + E_m = E - IR_g$$

According to *Jacobi's law*, maximum motor power and torque occur when $E_m = IR_m$, or when the back emf equals one-half the motor terminal voltage. Mechanical power output is equal to the product of armature current and back emf, or:

$$P_{\text{mech}} = IE_m$$

Therefore:

$$P_{\text{mech max}} = IV_t/2$$

In order to express power in terms of source voltage E, redraw the circuit so that $V_t \triangleq E$:

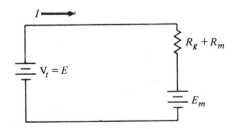

Now,

$$P_{mech\ max} = IE/2$$

$$E_m = I(R_g + R_m) = \frac{E}{2}, \quad I = \frac{E}{2(R_g + R_m)}$$

Finally,

(answer)

$$P_{mech\ max} = \frac{E^2}{4(R_g + R_m)}$$

Determining motor speed:

$$E_m = \frac{E}{2} = I(R_g + R_m) = K\omega I$$

(answer)

$$\omega = \frac{(R_g + R_m)}{K}$$

If your answers are correct, go on to Problem 4-2.
If your answers are not correct, review above solution.

PROBLEM 4-2. SHUNT-WOUND DC MOTOR

For the shunt-wound dc motor having the circuit parameters shown in the schematic below, determine the armature current and field current when its speed drops to 1100 rpm under load. No-load speed is nearly 1200 rpm at an armature current of 1 amp.

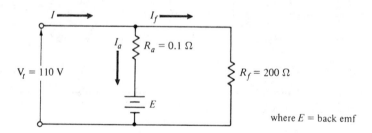

Solution:

$$V_t = I_a R_a + E$$

$$E = V_t - I_a R_a = 110 - 0.1 I_a$$

Since E is proportional to speed,

$$\frac{E_l}{E_{nl}} = \frac{1100}{1200} = \frac{110 - 0.1 I_{al}}{110 - 0.1 I_{anl}}$$

$$I_{al} = -\frac{1100}{1200}\left[\frac{110 - 0.1 I_{anl}}{0.1}\right] + \frac{110}{0.1}, \quad I_{anl} = 1 \text{ amp}$$

$\therefore I_{al} = 92.58$ amps (answer)

$$I_f = \frac{V_t}{R_f} = \frac{110}{200} = 0.55 \text{ amp}$$ (answer)

If your answers are correct, go on to Problem 4-3.
If your answers are not correct, review pp. 92 and 93.

PROBLEM 4-3. COMPOUND (SERIES-SHUNT)-WOUND DC MOTOR

For the series–shunt-wound dc motor having the circuit parameters shown in the schematic *low* on page 70, find the line current and gross mechanical hp output when the armature current $I_a = 25$ amps.

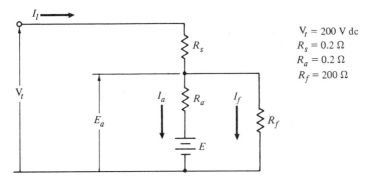

$V_t = 200$ V dc
$R_s = 0.2\ \Omega$
$R_a = 0.2\ \Omega$
$R_f = 200\ \Omega$

Solution:

$$I_f = \frac{E_a}{R_f} = \frac{200 - I_l R_s}{R_f} = \frac{200 - 0.2 I_l}{200} = 1 - 0.001 I_l$$

$$I_l = I_a + I_f = 25 + I_f = 25 + 1 - 0.001 I_l = 26 - 0.001 I_l$$

$$I_l = \frac{26}{1.001} = 25.974 \text{ amps}$$ (answer)

Gross mechanical hp $= I_a E / 746$

$$E_a = I_a R_a + E = V_t - I_l R_s = 200 - 25.97 \times 0.2 = 194.81 \text{ volts}$$

$$E = 194.81 - I_a R_a = 194.81 - 25 \times 0.2 = 189.81 \text{ volts}$$

$$HP = \frac{25 \times 189.81}{746} = 6.361 \text{ hp}$$ (answer)

If your answers are correct, go on to Problem 4-4.
If your answers are not correct, review p. 92.

PROBLEM 4-4. SHUNT MOTOR TORQUE

A dc shunt wound motor has the following characteristics:

$R_a = 0.5\ \Omega$ no-load speed = 1800 rpm
$R_f = 200\ \Omega$ HP = 7.5

If the no-load current is 3.5 amps when applied voltage is 230 volts, find the line current and developed torque when the motor is loaded such that the speed drops to 1700 rpm.

Solution:

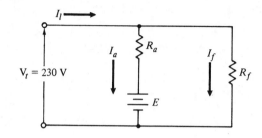

$$I_f = \frac{V_t}{R_f} = \frac{230}{200} = 1.15 \text{ amps}$$

$$I_a = I_l - I_f = 3.5 - 1.15 = 2.35 \text{ amps @ 1800 rpm}$$

Speed in rpm:

$$S = \frac{V_a - I_a R_a}{K} = \frac{E}{K}$$

where

K is a constant
V_a is the voltage across the armature

Solving for K:

$$K = \frac{230 - 2.35 \times 0.5}{1800} = 0.127125$$

Solving the preceding equation for I_a yields:

$$I_a = \frac{V_a - KS}{R_a}$$

At 1700 rpm:

$$I_a = \frac{230 - 0.127125 \times 1700}{0.5} = 27.775 \text{ amps}$$

(answer) $$I_l = I_a + I_f = 28.925 \text{ amps}$$

Developed torque is defined by the formula:

$$T_d = 7.04 \, KI_a = 7.04 \times 0.127125 \times 27.78$$

$$= 24.86 \text{ lb-ft} \qquad \text{(answer)}$$

Solving for torque by an alternative method:

$$HP = \frac{ST}{5252.12}$$

Rewriting:

$$T = \frac{5252.12 \, HP}{S}$$

where

$$HP = \frac{I_a E}{746}$$

$$E = 230 - 27.78 \times 0.5 = 216.11$$

$$\therefore T = \frac{5252.12}{1700} \left[\frac{27.78 \times 216.11}{746} \right] = 24.86 \text{ lb-ft} \qquad \text{(answer)}$$

If your answers are correct, go on to Problem 4-5.
If your answers are not correct, review p. 92.

PROBLEM 4-5. SEPARATELY-EXCITED DC GENERATOR

A separately-excited 220 volt dc generator has a 5300 watt rating at 3000 rpm for an input power of 8 HP. Armature resistance is 0.4 ohms. It is proposed to increase the output voltage of this machine to 260 V_{dc} at a speed of 3600 rpm.

Determine the machine ratings at the new operating conditions. Also determine the field current if the generator voltage is $E_g = 100 \, I_f - I_f^2$.

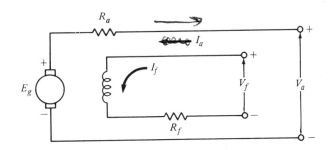

Solution:

Generated voltage and power are proportional to speed.

Therefore:

(answer)
$$V_a \text{ @ 3600 rpm} = 200 \times \frac{3600}{3000} = 264 \ V_{dc}$$

(answer)
$$P_{out} \text{ @ 3600 rpm} = 5.3 \text{ KW} \times \frac{3600}{3000} = 6.36 \text{ KW}$$

Input horsepower also increases with speed, but not in exact proportion because part of the input power must supply $I_a^2 R_a$ losses, which do not vary with speed.

$$I_{full \ load} = \frac{5.3 \text{ KW}}{220 \text{ volts}} = 24.09 \text{ amps}$$

$$I_a^2 R_a = 24.09^2 \times 0.4 = 232.15 \text{ watts or } 0.31 \text{ HP}$$

Assuming the remainder of the 8 HP input is due to torque–speed, the speed increase will raise this portion of the input to:

$$(8 - 0.31) \times \frac{3600}{3000} = 9.23 \text{ HP}$$

The total input HP, then, is:

$$9.23 + 0.31 = 9.54 \text{ HP} \qquad \text{(answer)}$$

At full load for a terminal voltage of 260 V_{dc} and a full load current of 24.09 amps through an armature resistance of 0.4 ohms, the generated voltage must be:

$$E_g = 260 + 24.09 \times 0.4 = 269.64$$

Calculating the required field current:

$$269.64 = 100 \ I_f - I_f^2$$

$$I_f^2 - 100 \ I_f + 269.64 = 0$$

$$I_f = 2.77 \text{ amps} \qquad \text{(answer)}$$

If your answers are correct, go on to the next section.
If your answers are not correct, review pp. 93 and 94.

AC MACHINES

You may wish to return to Table 4-1 (p. 65) and review the classification chart of electric motors. There are many more variables to consider when dealing with ac machines.

Synchronous Generators

Synchronous generators do not differ in principle from dc generators. In fact, any dc generator is a synchronous generator in which the alternating voltage set up in the armature inductors is rectified by means of a commutator. Although any dc generator may be used as a synchronous generator by the addition of *slip rings* electrically connected to suitable points of its armature winding, it is more satisfactory to interchange the moving and fixed parts when only ac currents are to be generated. The only moving parts required are those necessary for the field excitation, which is carried at low potential.

The frequency of any synchronous generator is given by the formula:

$$f = \frac{p}{2} \times \frac{n}{60} = \frac{pn}{120}$$

where

$p/2$ is the number of pole pairs, and
n is speed in rpm.

The voltage induced in a dc generator is equal to the average voltage induced in the coils in series between the brushes. In a synchronous generator, the induced voltage is the effective, or RMS, voltage. The instantaneous voltage induced in any coil on the armature of an alternator is:

$$e = -N\frac{d\phi}{dt}$$

where

N is the number of turns in the coil, and
ϕ is the flux enclosed by the coil at any instant.

It can be shown that the rms voltage per coil is:

$$E = 4.44 N f \phi_m \text{ volts}$$

when flux distribution in the air gap is such as to produce a sine wave in the armature coils.

The *rating* or maximum output of any synchronous generator is limited by its mechanical strength, by the temperature rise of its parts produced by losses, and by the permissible increase in field excitation necessary to maintain rated voltage at some specified load and power factor. The maximum voltage at a given frequency depends upon the permissible pole flux and field heating. Synchronous generators are rated in KVA at a given frequency or in KW at a stated power factor, frequency, and voltage.

The *regulation* of a synchronous generator at a given power factor is the percentage rise in voltage under conditions of constant excitation and frequency when the rated KVA load at a given power factor is removed. Regulation is positive for an inductive load, since the voltage rises when the load is removed; conversely, regulation is negative for a capacitive load if the load angle is sufficiently great for the voltage to fall.

Efficiency of a synchronous generator is defined as

$$\frac{\text{output}}{\text{output + losses}}$$

i.e., the same as for dc generators. The standard conditions under which efficiency is determined are rated voltage, frequency, power factor, and rated load, at 75°C. Efficiency is usually determined by output and losses. Losses are copper losses in the armature and field windings, core losses, and mechanical losses.

When two or more generators are operated in parallel, they must run at exactly the same frequency, phase, and voltage. When synchronous generators are paralleled, it is necessary to synchronize them before closing the switch. Once connected, the natural reactions resulting from departure from synchronism tend to re-establish synchronism. Unless mechanically coupled, synchronous generators cannot ordinarily operate in series. They are stable only in parallel.

Synchronous Motors

There is no essential difference in construction between synchronous motors and generators. Any synchronous generator can operate as a synchronous motor and vice versa. As a general rule, a synchronous motor is designed with a more effective damping device to prevent hunting, and its armature reaction is larger than is desirable for a generator. A synchronous motor operates at one speed, the synchronous speed, which depends solely upon the number of poles and the frequency of the line voltage. Speed is independent of load. A load change is accompanied by a change in phase and in instantaneous speed but not by a change in average speed. If, because of excessive load or any other cause, the average speed differs from synchronous speed, the average torque developed is zero, and the motor comes to rest. A synchronous motor, as such, has no starting torque.

The power factor of a synchronous motor operating from a constant potential is fixed by its field excitation and load. At any given load the power factor can be varied over wide limits by altering the field excitation. A motor may be overexcited, underexcited, or normally excited. Normal excitation produces unity power factor. Overexcitation produces capacitive action and causes the motor to take leading current. Underexcitation causes the motor to take lagging current. The field current which produces normal excitation depends upon the load and, in general, it increases with the load.

A synchronous motor operating with constant impressed voltage and frequency can operate at constant speed only under constant load conditions. The speed in rpm is given by the formula:

$$n = \frac{2f}{p} \times 60$$

This is another version of the formula given previously for the synchronous generator. As mentioned above, if a load is increased, the motor begins to slow down and continues to do so until sufficient change in phase has been produced. If the motor is not properly damped, it may overrun and develop too much power; it then speeds up and may again overrun in the reverse direction, developing too little power. Repeated action of this sort is called *hunting*.

A synchronous motor is not inherently self-starting. The average synchronous motor torque is zero at rest and until synchronous speed is reached. Some auxiliary device is necessary to bring a synchronous motor up to speed.

There are many uses for synchronous motors. A synchronous motor is frequently used for one unit of a motor–generator set when the other unit is a dc generator or a synchronous generator. When the other unit is a synchronous generator, the M–G set acts as a frequency converter (e.g., 50 to 60 Hz or vice versa). When operated at unity PF, a synchronous motor weighs and costs less than an equivalent induction motor.

Synchronous motors are often operated without load on a power transmission system to control power factor and to improve voltage regulation. A polyphase synchronous motor floated on a circuit carrying an unbalanced load tends to restore balanced conditions in regard to both current and voltage. If the system is badly out of balance, the synchronous motor may take power from the phases at high voltage and deliver power to the phase, or phases, at low voltage.

Asynchronous Machines

The previous paragraphs considered machines which operate at synchronous speeds. There is another class known as asynchronous machines which do not operate at synchronous speed. Their speed varies with the load.

A commercial synchronous machine requires fields excited by direct current. An asynchronous machine requires no dc excitation. Both parts of an asynchronous machine (armature and field) carry alternating current and are either connected in series, as in the series motor, or are inductively related, as in the induction motor. The induction motor and generator, the series motor, the repulsion motor, and all forms of ac commutator motors are included in the general class known as asynchronous machines. The induction motor is the most important and most widely used type of asynchronous motor. It has essentially the same speed and torque characteristics as a dc shunt motor and is suitable for the same type of work.

Polyphase Induction Motor

The polyphase induction motor is equivalent to a static transformer operating on a noninductive load whose magnetic circuit is separated by an air gap into two portions which rotate with respect to each other. In most cases the primary is the stator and the secondary is the rotor. The stator winding is similar to the armature winding of a synchronous machine. The polyphase currents in the stator winding produce a rotating magnetic field corresponding to the armature reaction in a synchronous generator. The fundamental of this field revolves at synchronous speed with respect to the stator in the same way the armature reaction of a synchronous machine revolves with respect to the armature. With respect to the rotor, the field revolves at a speed which is the difference between synchronous speed and the speed of the rotor. This difference is known as *slip*.

Slip is defined by the formula:

$$S = \frac{n_1 - n_2}{n_1}$$

where

n_1 is the speed of the revolving magnetic field and also the synchronous speed in rpm, and

n_2 is the actual speed of the rotor in rpm.

The formula for n_1 is:

$$n_1 = \frac{120 f_1}{p}$$

Induction Generator

An induction generator does not differ in construction from an induction motor. Whether an induction machine acts as a generator or as a motor depends solely upon its slip. Below synchronous speed it can operate only as a motor; above synchronous speed it operates as a generator.

The power factor at which an induction generator operates is fixed by its slip and its constants and not by the load.

An induction generator is free from hunting since it does not operate at synchronous speed.

Single-Phase Induction Motors

A single-phase induction motor possesses no starting torque. It is heavier, has lower efficiency and lower power factor than a polyphase motor for the same speed and output. These characteristics are true for any single-phase motor or generator.

The single-phase induction motor must be started by some form of auxiliary device and must attain considerable speed before it develops sufficient torque to overcome its own friction and windage. The direction of rotation depends merely upon the direction in which it is started. Once started, it operates as well in one direction as the other.

PROBLEM 4-6. INDUCTION MOTOR SPEED

A three-phase induction motor has the following nameplate data:

$$15 \text{ hp, } 230 \text{ V, } 60 \text{ Hz, } 1150 \text{ rpm}$$

Determine the number of poles, slip and rotor current frequency at full load. Also, estimate the speed at half of full load.

Solution:

The curve shown below is a typical speed–torque motor curve:

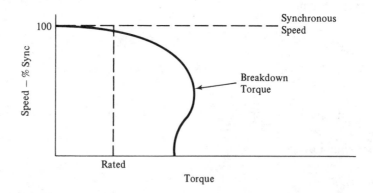

An induction motor operating at rated load runs at a speed somewhat lower than synchronous speed. In this case, synchronous speed is most likely 1200 rpm. Based on this assumption, the number of poles may be calculated as follows:

$$f = \frac{p}{2} \times \frac{n}{60}$$

where

f = frequency in Hz
$p/2$ = number of pole pairs
n = synchronous speed in rpm

Thus,

$$p = \frac{120 f_1}{n_1} = \frac{120 \times 60}{1200} = 6 \text{ poles} \qquad \text{(answer)}$$

Slip is defined by the formula:

$$S = \frac{n_1 - n_2}{n_1}$$

where

S = slip
n_1 = synchronous speed in rpm
n_2 = actual rotor speed in rpm

Thus,

$$S = \frac{1200 - 1150}{1200} = 0.0417 \qquad \text{(answer)}$$

Rotor current frequency is given by the formula:

$$f_s = S \times f = 0.0417 \times 60 = 2.5 \text{ Hz} \qquad \text{(answer)}$$

As a point of information, rotor currents at frequency f_s produce a rotating *magneto-motive* force on the rotor. This mmf revolves at a speed n_s with respect to the rotor. This relative speed is the difference between synchronous speed and actual speed, and is given by the formula:

$$n_s = n_1 - n_2 = \frac{120 f_1 S}{p}$$

Assuming the speed-torque curve is linear over the range of operation, speed varies directly with load. Thus, the speed at half of full load is calculated as follows:

$$\frac{7.5 \text{ hp}}{15 \text{ hp}} = \frac{1200 - n_2}{1200 - 1150} = \frac{1200 - n_2}{50} = \frac{1}{2}$$
$$n_2 = 1200 - 25 = 1175 \text{ rpm} \qquad \text{(answer)}$$

If your answers are correct, go to Problem 4-7.
If your answers are not correct, review pp. 99 and 101.

PROBLEM 4-7. INDUCTION MOTOR EFFICIENCY

A six-pole, 220 volt, 3-phase induction motor has the following measured losses:

    stator copper loss = 0.6 KW
    friction, no-load core and windage losses = 1 KW

The motor runs at 1150 rpm when drawing 25 KW from the line. Find the motor efficiency and output horsepower under these conditions.

Solution:

The equivalent circuit of a polyphase induction motor is shown below. It has been simplified so that there is a nearly constant error of 2–3% in induced voltages E_1 and E_2 between no load and full load. Everything in the equivalent circuit is referred to the primary.

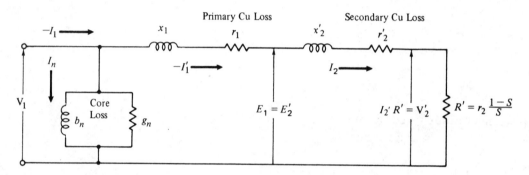

Losses in an induction motor include:

1. primary copper loss;
2. secondary copper loss;
3. core loss;
4. friction and windage loss;
5. stray-load loss.

When it is possible to load the motor and measure its input and slip but not possible to measure the output directly, the efficiency is obtained from the formula:

$$\eta = \frac{\text{input} - \text{losses}}{\text{input}}$$

Primary copper loss is usually obtained by measuring the ohmic resistance per phase with dc and multiplying by the square of the primary phase current and the number of phases. In this problem, this loss is given as:

$$0.6 \text{ KW} = 3I_1^2 r_1$$

Whenever slip is accurately measurable, the secondary copper loss may be taken as the product of the measured primary input minus the primary copper and core losses and the slip. In this case,

Note: The term "copper loss" is no longer accurate since many machines make use of aluminum rather than copper; "winding loss" is a better term.

$$\text{slip} = \frac{1200 - 1150}{1200} = 0.0417 = \frac{\text{sec. Cu loss}}{P_{\text{in}} - (\text{prim. Cu loss} + \text{core loss})}$$

Thus, rotor (secondary) copper loss = (25 - 0.6) 0.00417 = 1.017 KW.

Sometimes it is difficult to separate core loss from friction and windage losses. In this case, core loss may be excluded from the primary copper losses with little effect on the final result.

The friction and windage loss plus the core loss may be obtained by running the motor at no load with normal frequency and voltage applied. The input is then equal to the no-load primary copper loss, friction and windage, core loss and the no-load secondary copper loss (which is small and usually neglected). In this problem, these losses are given as 1 KW.

Stray-load loss includes all losses not otherwise accounted for. This is obtained from two tests: a dc excitation test and a test to find the blocked rotor torque with balanced polyphase voltages at rated frequency. In this problem, this loss will be assumed to be combined with all other losses.

Total losses, then, equal:

$$P_{\text{total loss}} = 0.6 + 1.017 + 1 = 2.617 \text{ KW}$$

Efficiency is:

$$\eta = \frac{25 - 2.617}{25} = 0.895 \text{ or } 89.5\%$$ (answer)

Output horsepower is:

$$P_{\text{out}} = \frac{25 - 2.617}{0.746} = 30 \text{ hp}$$ (answer)

If your answers are correct, go to Problem 4-8.
If your answers are not correct, review p. 92.

PROBLEM 4-8. INDUCTION MOTOR LOSSES

A 4 hp, 4-pole, 60 Hz, 3-phase, wound-rotor motor draws 3670 watts from the input power line. The following losses have been measured:

primary copper (stator) loss: 350 watts
core loss: 200 watts
friction, windage, stray-load loss: 50 watts

What is the speed of the motor when delivering 3.95 hp?

Solution:

$$n_1 = \frac{120f}{4} = \frac{120 \times 60}{4} = 1800 \text{ rpm}$$

$$n_2 = n_1(1 - S) = 1800(1 - S)$$

105

$$S = \frac{\text{secondary copper loss}}{P_{\text{in}} - (\text{primary copper loss} + \text{core loss})} = \frac{\text{secondary copper loss}}{P_{em}}$$

$$= \frac{\text{secondary copper loss}}{3670 - 550}$$

We need to find secondary copper loss.

$$P_{\text{total losses}} = P_{\text{in}} - P_{\text{out}} = 3670 - 3.95 \times 746$$

$$= 3670 - 2946 = 723.3 \text{ watts}$$

Secondary copper losses $= P_{\text{total loss}} -$ all other losses

$$= 723.3 - (350 + 200 + 50) = 123.3 \text{ watts}$$

Now we can calculate slip from the above formula (assuming core loss charged to stator):

$$S = \frac{123.3}{3670 - 550} = 0.0395$$

Finally, the speed is:

(answer)
$$n_2 = 1800(1 - S) = 1800(1 - 0.0395) = 1729 \text{ rpm}$$

If core loss is charged to rotor,

$$S = \frac{123.3}{3670 - 350} = 0.0371$$

$$n_2 = 1800(1 - 0.0371) = 1733 \text{ rpm}$$

Another way to calculate slip is from the formula:

$$P_{\text{out}} = \text{rotor loss} \times \left(\frac{1 - S}{S}\right)$$

$$2946.7 = 123.3 \left(\frac{1 - S}{S}\right), \quad \frac{1 - S}{S} = 23.9$$

(answer)
$$1 - S = 23.9S, \quad 1 = 24.9S, \quad S = \frac{1}{24.9} = 0.0402$$

If your answers are correct, go to Problem 4-9.
If your answers are not correct, review pp. 102 and 103.

PROBLEM 4-9. INDUCTION MOTOR SPEED

Find the speed of a three-phase, 60 Hz, 12 HP, 6 pole induction motor, given the following information:

$P_{\text{in}} = 11,000$ watts
$P_{\text{out}} = 11.9$ HP

P_{core} = 540 watts

$P_{\text{stator copper loss}}$ = 770 watts

$P_{\text{rotational}}$ = 175 watts

Solution:

$$P_{\text{in}} - P_c - P_s = 11{,}000 - 540 - 770 = 9690 \text{ watts}$$

$$I_R{}^2 R_R = 9690\,S$$

$$P_{\text{out}} + P_{\text{rot}} = 11.9 \times 746 + 175 = 9052 \text{ watts}$$

$$I_R{}^2 R_R \frac{(1-S)}{S} = 9052 = 9690\,(1-S)$$

$$1 - S = \frac{9052}{9690} = 0.934$$

$$n = \frac{60f}{P/2} = \frac{3600}{6/2} = 1200 \text{ RPM}$$

$$n_r = n_s\,(1-S) = 1121 \text{ RPM} \qquad \text{(answer)}$$

If your answer is correct, go on to Problem 4-10.

If your answer is not correct, review the previous two problems.

PROBLEM 4-10. MOTOR STARTING-LINE VOLTAGE DROP

The impedance of a 3-phase power distribution system is $0.06 + j0.3$. Open circuit line voltage is 2400 volts. Calculate the percent voltage drop when a 400 hp induction motor is connected to the line and started.

Solution:

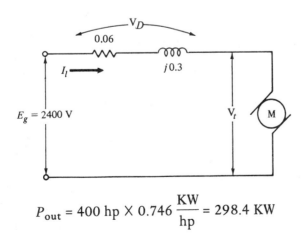

$$P_{\text{out}} = 400 \text{ hp} \times 0.746\,\frac{\text{KW}}{\text{hp}} = 298.4 \text{ KW}$$

Before this problem can be solved, certain required information not given must be assumed about the motor:

107

$$\eta = 89\%$$

$$PF_{FL} = 0.9 \text{ lag}$$

$$I_{start} = 6\,I_l$$

$$PF_{start} = 0.25 \text{ lag}$$

$$P_{out} = \sqrt{3}\, V_l I_l \eta \cos\theta$$

$$I_l = \frac{P}{\sqrt{3}\, V_l \eta \cos\theta} = \frac{298.4}{\sqrt{3}\,(2.4)(0.89)(0.9)} = 89.62 \text{ amps}$$

$$I_{start} = 6 I_l = 537.7 \text{ amps}$$

Using I as reference:

$$V_t = E_g - V_D = \frac{2400}{\sqrt{3}}\,(0.25 + j0.968) - 537.7(0.06 + j0.3)$$

$$= 1385.64\underline{/75.5°} - 164.5\underline{/78.7°} = 1221.4\underline{/75.1°}$$

(answer)
$$\% \text{ drop} = \frac{1385.64 - 1221.4}{1385.64} \times 100 = 11.85\%$$

An alternative way to calculate voltage drop of the distribution line is to use the following formula:

$$V_{drop} = IR \cos\theta + IX \sin\theta$$

$$= 537.7 \times 0.06 \times 0.25 + 537.7 \times 0.3 \times 0.968 = 164.22 \text{ volts}$$

Thus:

(answer)
$$\% \text{ drop} = \frac{164.22}{1385.66} \times 100 = 11.85\%$$

If your answers are correct, go to Problem 4-11.
If your answers are not correct, review pp. 67 and 82.

PROBLEM 4-11. INDUCTION MOTOR CONNECTIONS

The following test data was obtained on a 60 HP, 900 rpm, three-phase, 60 Hz, 440 volt, delta-connected induction motor:

	Voltage	Current	Power	Torque
No load	440 v	31 a	2.6 KW	0
Loaded rotor	230 v	202 a	30.0 KW	144 lb-ft

What will be the starting torque and current if the machine is reconnected in wye and 440 volts line-to-line is applied?

Solution:

If the line voltage remains at 440 and the windings are connected in wye, the new phase voltage is:

$$V_\phi = \frac{440}{\sqrt{3}} = 254 \text{ volts}$$

Phase current in the delta-connected configuration with a locked rotor line current of 202 amps was:

$$I_\phi = \frac{202}{\sqrt{3}} = 116.6 \text{ amps}$$

$$T = \frac{7.045}{n_1} EI_a$$

Torque is proportional to the square of the voltage as seen from the preceding equation, because voltage is proportional to current ($V = IR$). Calculating new starting torque based on the original locked rotor phase voltage vs. the new phase voltage of 254 volts:

$$T_{ST} = \left(\frac{254}{230}\right)^2 \times 144 = 176 \text{ lb-ft} \qquad \text{(answer)}$$

Since current is directly proportional to voltage, the new starting current per phase is:

$$I_{ST} = \frac{254}{230} \times 116.6 = 129 \text{ amps} \qquad \text{(answer)}$$

If your answers are correct, go on to the next section.
If your answers are not correct, review p. 92.

Two-Winding Transformers

A two-winding transformer consists of two or more insulated coils coupled by mutual induction. Its action depends upon the self-inductances of its coils and the mutual induction between them. In its simplest form, the transformer consists of two coils linked by a common magnetic circuit. When power is supplied to one coil at a definite frequency and voltage, power can be taken from the other at the same frequency, but, in general, at a different voltage.

Transformers may be of the air-core or iron-core types, but for power purposes they are always of the iron-core type.

The voltage induced in any winding depends only on the number of turns in the winding and the rate of change of the flux linking it. If N_1 is the number of turns in the coil, the voltage rise induced in the coil at any instant by the flux ϕ is:

$$e = -N_1 \frac{d\phi}{dt} = -\omega N_1 \phi_m \cos \omega t$$

The maximum voltage is:

$$e_{\max} = \omega N_1 \phi_{\max}$$

$$= 2\pi f N_1 \phi_{\max}$$

The effective RMS voltage in volts, when ϕ_{max} is expressed in webers, is:

$$E_1 = \frac{2\pi f}{\sqrt{2}} N_1 \phi_{max} = 4.44 f N_1 \phi_{max}$$

If the voltage is not a sine wave, then:

$$E_1 = 4 \times (\text{form factor})^* \times f N_1 \phi_{max}$$

The ratio of transformation of a transformer is given by the formula:

$$\frac{E_1}{E_2} = \frac{N_1}{N_2} = a$$

Efficiency of a transformer is defined as:

$$\eta = \frac{\text{full load output (watts) at rated PF}}{\text{full load output + losses}}$$

Losses include core loss and copper loss. These losses may be determined from open circuit (core loss) and short circuit (copper loss) tests.

PROBLEM 4-12. TRANSFORMER EFFICIENCY AND REGULATION

Short circuit and open circuit tests were run on a single-phase transformer to determine its characteristics. The 2400/240 volt, 60 Hz transformer is rated at 100 KVA. Test data is shown below:

1. High-voltage winding open; rated voltage applied to the low-voltage terminals:

$$I = 32 \text{ amps}$$
$$P = 600 \text{ watts}$$
$$V = 240 \text{ volts}$$

2. High-voltage winding short-circuited; voltage applied to the low-voltage winding such that rated current flows in the windings:

$$I = 417 \text{ amps}$$
$$P = 800 \text{ watts}$$
$$V = 15 \text{ volts}$$

Determine efficiency η and full load regulation at PF = 0.8 lag. Also, determine efficiency at 1.25 times rated load.

Solution:

The equivalent circuit of a two-winding transformer is shown on p. 111.

---

*Where the *form factor* is the ratio of the rms value of the waveform to its half-period average value, the form factor for a sine wave is 1.11.

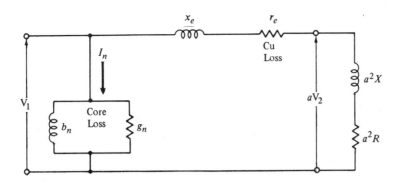

$$\left.\begin{aligned} x_e &= x_1 + a^2 x_2 \\ r_e &= r_1 + a^2 r_2 \end{aligned}\right\} \quad \text{referred to primary}$$

In this figure, x_e and r_e are the equivalent reactance and equivalent resistance referred to the primary side and must be used with the secondary current referred to the primary side. The open-circuit test (test 1) gives core loss data, and the short-circuit test (test 2) gives copper loss data.

Efficiency is given by the formula:

$$\eta = \frac{\text{full load output (watts) at rated PF}}{\text{full load output (watts) at rated PF} + \text{core loss} + \text{copper loss}}$$

Thus,

$$\eta = \frac{100{,}000 \times 0.8}{100{,}000 \times 0.8 + 600 + 800} = 0.9828 \text{ or } 98.28\% \qquad \text{(answer)}$$

$$\text{Regulation} = \frac{|V_{\text{no load}}| - |V_{\text{full load}}|}{|V_{\text{full load}}|} = \frac{|V_{\text{in}}| - |V_{\text{out}}|}{|V_{\text{out}}|}$$

$$\frac{V_1}{a} = V_{\text{in}} = V_2 + I_2 (\cos\theta \pm j \sin\theta)(r_e + jX_e)$$

$$\begin{cases} + \text{ for leading PF} \\ - \text{ for lagging PF} \end{cases}$$

The equivalent impedance is obtained from the short-circuit test:

$$P = I^2 r_e, \quad r_e = \frac{P}{I^2} = \frac{800}{417^2} = 0.00461 \ \Omega$$

$$Z_e = \frac{V}{I} = \frac{15}{417} = 0.036 = r_e + jx_e = 0.00461 + j0.0357$$

$$= 0.036\underline{/82.64°}$$

Calculate V_{in}:

$$V_{\text{in}} = 240 + 417(0.8 - j0.6)(0.036\underline{/82.64°}) = 250.7\underline{/2.46°}$$

Regulation may now be determined:

(answer)

$$\text{Regulation} = \frac{250.7 - 240}{240} = 0.0446 \text{ or } 4.46\%$$

Efficiency at 1.25 of full load is calculated as follows: copper loss varies as the square of the load; thus,

$$P_{in} = 800 \times 1.25^2 = 1250 \text{ watts}$$

At 125% full load:

$$P_{out} = 100,000 \times 0.8 \times 1.25 = 100,000 \text{ watts}$$

(answer)

$$\eta = \frac{100,000}{100,000 + 600 + 1250} = 0.9818 \text{ or } 98.18\%$$

If your answers are correct, go to Problem 4-13.
If your answers are not correct, review pp. 68, 82, 91, 107 and 108.

PROBLEM 4-13. REGULATION IMPROVEMENT

A 750 KVA, 3-phase, balanced load of 60% power factor is supplied from a 1000 KVA, 3-phase transformer and distribution line. The line impedance relative to the 1000 KVA is 3% resistance and 5% reactance. What will be the voltage regulation improvement if the power factor is corrected to 85%?

Solution:

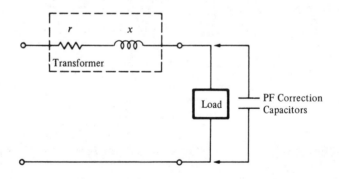

This problem is easily solved using the equation:

$$\% \text{ regulation} = 100 \left[mr + nx + \frac{(mx - nr)^2}{2} \right] \left[\frac{\text{actual KVA load}}{\text{rated KVA load}} \right]$$

where

$m = $ PF of load $= \cos \theta$

$n = \pm \sin \theta (+ \text{ for lag}, - \text{ for lead})$

$r = $ resistance factor $= \dfrac{\text{resistance loss in KW}}{\text{rated KVA of transformer}}$

x = reactance factor = $\sqrt{Z^2 - r^2}$

$$Z = \frac{\text{impedance in KVA}}{\text{rated KVA of transformer}}$$

The terms r and X are on a per-unit basis. Thus, if resistance and reactance factors are 3% and 5%, then $r = 0.03$ and $X = 0.05$, respectively.

Before PF correction, the phasor diagram is:

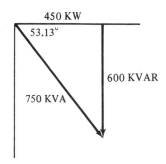

$$\% \text{ Regulation} = 100 \left[0.6 \times 0.03 + 0.8 \times 0.05 + \frac{(0.6 \times 0.05 - 0.8 \times 0.03)^2}{2} \right] \frac{750}{1000}$$

$$= 100 \, [0.018 + 0.04 + 0.000018] \, 0.75 = 4.35\%$$

After PF correction, the phasor diagram becomes:

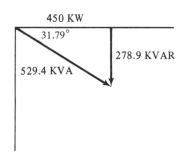

$$\% \text{ regulation} = 100 \left[0.85 \times 0.03 + 0.527 \times 0.05 + \frac{(0.85 \times 0.05 - 0.527 \times 0.03)^2}{2} \right] \frac{529.4}{1000}$$

$$= 100 \, [0.0255 + 0.02635 + 0.0004] \, 0.5294 = 2.75\%$$

Regulation improvement = 4.35 – 2.75 = 1.6%   (answer)

The number of KVARs of capacitance to achieve this improvement is:

600 – 279 = 321 KVARS

If your answers are correct, go to Problem 4-14.
If your answers are not correct, review p. 67.

PROBLEM 4-14. TRANSFORMER SPECIFICATIONS

A new machine shop is being established and will have the following loads:
   (1) 55 KVA, 240 volt, single-phase, unity power factor
   (2) 14 KW, 240 volt, three-phase, 0.82 power factor
Primary power to the shop is 7200 volts, three phase, 60 Hz. Design a power distribution system for these two loads.

Solution:

With the single-phase load considerably larger than the three-phase load, a V-V, or open delta, transformer configuration would be the most economical to install. Assuming an *a-b-c* phase sequence and the single-phase load across line *a-b* as follows, the ratings of the two transformers may be specified as follows:

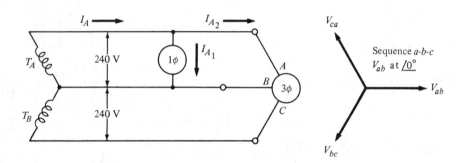

$$I_{A1} = I_{1\phi} = \frac{KVA}{V_{ab}} = \frac{55,000}{240 \underline{/0^\circ}} = 229.17 \underline{/0^\circ} \text{ amps}$$

$$I_{A2} = I_{3\phi} = \frac{P}{\sqrt{3}\, V_L \cos\theta} \underline{/-(\theta + 30^\circ)} = \frac{14,000}{\sqrt{3} \times 240 \times 0.82} \underline{/-61.79}$$

$$= 41.07 \underline{/-61.79^\circ} = 19.41 - j36.19 \text{ amps}$$

$$I_A = I_{A1} + I_{A2} = 229.17 + 19.41 - j36.19 = 248.58 - j36.19 = 251.2 \underline{/-8.28} \text{ amps}$$

$$KVA_A = (KV)(I_A) = 0.24 \times 251.2 = 60 \text{ KVA}$$

$$KVA_B = (KV)(I_{3\phi}) = 0.24 \times 41.07 = 9.86 \text{ KVA}$$

In summary, the two transformers to be installed have the following specifications:

(answer) T_A: 60 KVA, 7200-to-120/240 volts, 60 Hz, 1ϕ
(answer) T_B: 10 KVA, 7200-to-120/240 volts, 60 Hz, 1ϕ

If your answers are correct, go on the next section.
If your answers are not correct, review p. 68.

Autotransformer

In many cases either an autotransformer or a two-winding transformer may be used to accomplish the same transformation. In such a case, they would have the same primary and

114

secondary voltages and equal KVA ratings, which would mean also the same rated values of primary and secondary currents. The size of wire used for the continuous winding is not the same throughout unless the ratio of transformation is such that its two parts carry equal currents. Since part of the winding serves for both primary and secondary, an autotransformer requires less material and is therefore cheaper than a two-winding transformer of the same output and efficiency. The saving, however, is substantial only when the ratio of transformation is small (less than 2).

The advantages of the autotransformer are better regulation and efficiency and lower cost for the same output. Its disadvantages are that its low-voltage winding is in electrical connection with the high-voltage winding and forms part of it, and that because of its reduced impedances larger currents result from a short-circuit on its secondary side.

For purposes of comparison, the following assumptions are made with respect to the autotransformer:

1. Its core is identical with that of the corresponding two-winding transformer.
2. The number of turns used in its winding equals that of the high-voltage winding transformer. The two transformers would then operate at the same flux density and their core losses would be equal.
3. The wire size for the series winding is the same as that used in the high voltage winding of a two-winding transformer. At rated load these windings carry equal currents.
4. The wire size for the common winding is so chosen that at rated load the current density in this winding equals the current density in the low voltage winding of the two-winding transformer at the same load.

A USEFUL FORMULA FOR DETERMINING THE RATING OF THE SERIES OR COMMON WINDING IS: $\text{RATING} = \text{OUTPUT}\left[1 - \dfrac{V_2}{V_1}\right]$

PROBLEM 4-15. AUTOTRANSFORMER CURRENTS

An autotransformer is adjusted to provide 240 volts to a 5 KVA single-phase load from a 480 volt distribution line. Calculate currents in the series and common winding of the auto-transformer.

Solution:

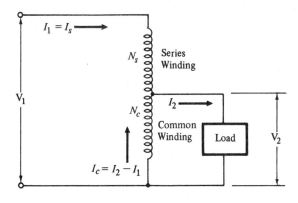

The turns ratio is:

$$a = \frac{V_1}{V_2} = \frac{N_1}{N_2} = \frac{480}{240} = 2$$

where

$$N_1 = N_s + N_c$$

$$N_2 = N_c$$

Assuming an ideal transformer (no losses):

(answer)

$$I_1 = \frac{5,000}{480} = 10.42 \text{ amps} = I_S$$

$$I_2 = \frac{5,000}{240} = 20.83 \text{ amps}$$

(answer)

$$I_c = I_2 - I_1 = 20.83 - 10.42 = 10.42 \text{ amps}$$

If your answers are correct, go to Problem 4-16.
If your answers are not correct, review pp. 114 and 115.

PROBLEM 4-16. AUTOTRANSFORMER RATING

A 220 microfarad capacitor is connected across the primary of a 440/120 volt autotransformer as follows:

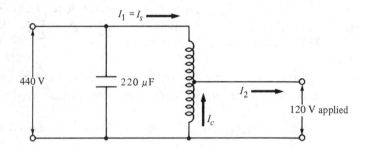

What is the equivalent capacitance when referred to the secondary? What are the KVA ratings of the common and series windings when 120 volts ac is impressed across the secondary winding?

Solution:

The turns ratio is:

$$a = \frac{440}{120} = 3.67$$

Calculating the equivalent secondary capacitance:

$$X_{C_p} = a^2 X_{Cs}, \quad \frac{1}{\omega C_p} = \frac{a^2}{\omega C_s}$$

(answer)

$$C_s = a^2 C_p = (3.67)^2 (220 \ \mu f) = 2958 \ \mu f$$

116

Calculating the winding KVAs:

$$X_{C_p} = \frac{1}{\omega_c C_p} = \frac{1}{377 \times 220 \times 10^{-6}} = 12.06 \ \Omega$$

$$I_s = \frac{440}{12.06} = 36.49 \ \text{amps}$$

$$I_2 = aI_s = 3.67 \times 36.49 = 133.81 \ \text{amps}$$

$$I_c = I_2 - I_s = 133.81 - 36.49 = 97.32 \ \text{amps}$$

$$(\text{KVA})_{\text{series}} = (\text{KV})_{\text{series}} I_s = (0.440 - 0.120) \times 36.49 = 11.68 \ \text{KVA} \qquad \text{(answer)}$$

$$(\text{KVA})_{\text{common}} = (\text{KV})_{\text{common}} I_c = (0.120) \times 97.32 = 11.68 \ \text{KVA} \qquad \text{(answer)}$$

An alternate method of calculating the rating of either the common or series winding of an autotransformer (they are always the same) is to use this formula:

$$\text{rating} = \text{output} \left[1 - \frac{V_2}{V_1}\right]$$

$$\text{rating} = 440 \times 36.49 \left[1 - \frac{120}{440}\right] = 16.06 \times 10^3 \ [0.73] = 11.68 \ \text{KVA} \qquad \text{(answer)}$$

If your answers are correct, go to the next section.
If your answers are not correct, review p. 115.

MAGNETIC DEVICES

The following are problems related to magnetic devices. The principles of magnetic circuits as discussed in Chapter 1 are applied to the solution of these problems.

PROBLEM 4-17. REACTOR

A reactive device has 250 turns of wire on a closed iron core having an average length of 18 inches. The iron cross-sectional area is 8 square inches. The device is to have an air gap cut in the iron so that the coil emf will be 120 volts RMS when the magnetizing current is 2.5 amps at 60 Hz.

Calculate the required length of the air gap.

Solution:

In order to solve this problem certain assumptions must be made. The following are reasonable assumptions:

1. effective cross-sectional air gap area including fringing = 12 square inches;
2. flux is uniformly distributed;

3. 12% of the total mmf is required to overcome the iron path reluctance;
4. winding factor = 1.

$$E_{rms} = 4.44 f N \phi_{max}$$

$$\phi_{max} = \frac{E_{rms}}{4.44 N f} = \frac{120}{4.44 \times 250 \times 60} = 1.802 \times 10^{-3} \text{ weber}$$

$$A_{air} = (12 \text{ in}^2)(6.45 \times 10^{-4} \text{ m}^2/\text{in}^2) = 7.74 \times 10^{-3} \text{ m}^2$$

$$B_{air} = \frac{\phi_{max}}{A_{air}} = \frac{1.802 \times 10^{-3} \text{ weber}}{7.74 \times 10^{-3} \text{ m}^2} = 0.233 \text{ weber/m}^2$$

$$l_{air} = \frac{\mu_0 \times 0.88 \, (NI_{max})_{air}}{B_{air}} \text{ meter}$$

$$I_{max} = \sqrt{2} I_{rms} = \sqrt{2} \times 2.5 = 3.54 \text{ amps}$$

(answer)
$$l_{air} = \frac{4\pi \times 10^{-7} \times 0.88 \times 250 \times 3.54}{0.233} = 4.2 \times 10^{-3} \text{ m}$$

If your answer is correct, go to Problem 4-18.
If your answer is not correct, review p. 21.

PROBLEM 4-18.   MAGNETIC DEVICE

A magnetic device is designed to operate at 115 V, 60 Hz with an eddy-current loss of 4 watts and a hysteresis loss of 9 watts. Find the core loss of the device when it is operated at 100 V, 50 Hz.

Solution:

$$\text{Core loss} = P_e + P_h$$

$$P_e = K_e t^2 f^2 B_{max}^2$$

$$P_h = \eta f B_{max}^n \quad \text{(Steinmetz empirical equation)}$$

where

K_e is coefficient of eddy-current loss
t is thickness of laminations
f is frequency
B_{max} is maximum flux density
η is coefficient of hysteresis loss
n is hysteresis exponent

$E = 4.44 f N \phi_{max}$, ϕ_{max} is directly proportional to B_{max}. Thus, B is directly proportional to the line voltage and inversely proportional to frequency.

$$\frac{B_{50}}{B_{60}} = \left(\frac{100}{115}\right)\left(\frac{60}{50}\right) = 1.04$$

$$P_{e_{50}} = P_{e_{60}} \left(\frac{50}{60}\right)^2 (1.04)^2 = 4 \times 0.76 = 3.02 \text{ watts}$$

$$P_{h_{50}} = P_{h_{60}} \left(\frac{50}{60}\right) (1.04)^{1.6} = 9 \times 0.89 = 8.03 \text{ watts}$$

Core loss$_{50}$ = 3.02 + 8.03 = 11.05 watts   (answer)

If your answer is correct, go to the next chapter.
If your answer is not correct, review above solution.

5 Control Theory

Discussions of and problems related to singularity functions, second-order systems, poles and zeros, partial fractions, Laplace transforms, stability, transfer function, compensation, Bode analysis, and root locus.

INTRODUCTION

The field of control theory brings together a broad spectrum of engineering principles and specialized mathematical and graphical techniques. This chapter presents the minimal technical background needed to demonstrate the solutions to the types of problems often found in the PE Exam. The solutions require a knowledge of basic feedback system terms, block diagram algebra, singularity functions, second order systems, differential equations, convolution, translation, inversion, Laplace transforms, pole–zero maps, *RLC* circuit analysis, partial fractions, stability, errors, system type, transfer functions, compensation, Bode analysis, root locus analysis, and system analysis and design.

If you have sufficient background in these areas, you will recognize the problem-solving techniques. If you do *not* have sufficient background, you may wish to study a basic college text before working through this chapter.

BASIC FEEDBACK SYSTEMS TERMS

A simple basic feedback system is illustrated as follows:

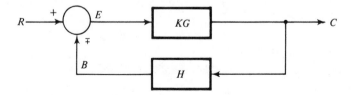

In the equations below, a minus sign denotes positive feedback; a plus sign denotes negative feedback.

K = constant forward loop gain

G = frequency-sensitive forward loop gain

E = error signal = $R - B$

R = reference or command input

$B = CH$ = primary feedback signal

C = controlled variable output

H = feedback gain ($H = 1$ for unity feedback)

KG = forward loop transfer function = forward gain

KGH = open loop transfer function

$$\frac{C}{R} = \frac{KG}{1 \pm KGH} = \text{system transfer function = closed loop gain}$$

$$\frac{B}{R} = \frac{KGH}{1 \pm KGH} = \text{primary feedback ratio}$$

$$\frac{E}{R} = \text{error ratio}$$

The *characteristic equation* of a feedback system is determined from $1 \pm KGH = 0$; alternatively, from the loop gain KGH, it is:

$$D_{GH} \pm N_{GH} = 0$$

where

D_{GH} is the denominator of KGH
N_{GH} is the numerator of KGH.

Examples:

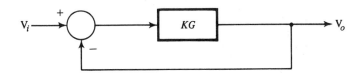

Unity negative feedback:

$$\frac{C}{R} = \frac{V_o}{V_i} = \frac{KG}{1 + KG}$$

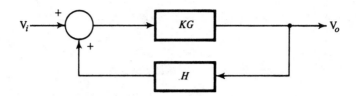

Positive feedback oscillator (seldom used):

$$\frac{C}{R} = \frac{V_o}{V_i} = \frac{KG}{1 - KGH}$$

SINGULARITY FUNCTIONS

Three *singularity functions* are used extensively in the study of control systems: the unit step, the unit impulse, and the unit ramp.

Unit Step, *u(t)*

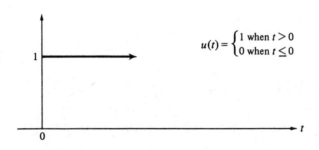

$$u(t) = \begin{cases} 1 \text{ when } t > 0 \\ 0 \text{ when } t \leq 0 \end{cases}$$

Response of a control system to the unit step input is the output $y(t)$ when input $x(t) = u(t)$ and all initial conditions are zero. Figure 5-1 shows the response of a second-order system to a unit step input for various damping factors δ. The integral of the unit impulse is the unit step.

Unit Impulse, *δ(t)*

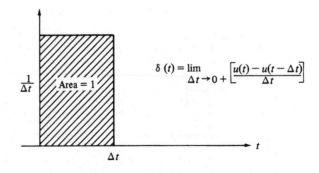

$$\delta(t) = \lim_{\Delta t \to 0+} \left[\frac{u(t) - u(t - \Delta t)}{\Delta t} \right]$$

The quotient in brackets represents a rectangle of height $1/\Delta t$ and width Δt. In the limit, the unit impulse is a rectangle of infinite height and zero width. The area under the curve is always equal to unity. The derivative of the unit step is the unit impulse.

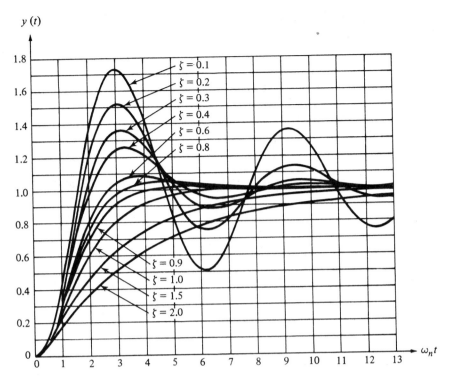

Figure 5-1. Universal transient response curves for second-order systems subjected to a unit step input.

Unit Ramp

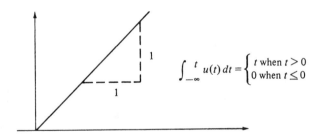

$$\int_{-\infty}^{t} u(t)\, dt = \begin{cases} t \text{ when } t > 0 \\ 0 \text{ when } t \leq 0 \end{cases}$$

The derivative of the unit ramp is the unit step. The integral of the unit step is the unit ramp.

SECOND-ORDER SYSTEMS

Second-order systems are important because most higher-order systems can be approximated by them. The general form of a linear, constant-coefficient, second-order differential equation is given below:

$$\frac{d^2 y}{dt^2} + 2\zeta\omega_n \frac{dy}{dt} + \omega_n^2 y = \omega_n^2 x$$

where

ζ is the *damping ratio* (*damping factor*) and
ω_n is the *undamped natural frequency* of the system.

123

The characteristic equation for the above equation is:

$$D^2 + 2\zeta\omega_n D + \omega_n^2$$

PROBLEM 5-1. SECOND-ORDER SYSTEM

The block diagram of a second-order servo system is shown below.

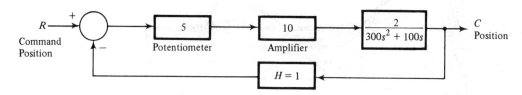

Determine the natural frequency and the damping ratio.

Solution:

The system transfer function for $H = 1$ is:

$$\frac{C}{R} = \frac{KG}{1 + KG} = \frac{1}{1 + \dfrac{1}{KG}} = \frac{1}{1 + \dfrac{300s^2 + 100s}{5 \times 10 \times 2}}$$

$$= \frac{100}{300s^2 + 100s + 100} = \frac{0.333}{s^2 + 0.333s + 0.333} = \frac{\omega_n^2}{s^2 + 2\zeta\omega_n s + \omega_n^2}$$

(answer)
$$\omega_n = \sqrt{0.333} \text{ rad/sec}$$

(answer)
$$\zeta = \frac{0.333}{2\sqrt{0.333}} = 0.29 \text{ (underdamped)}$$

If your answers are correct go to the next section.
If your answers are not correct, review pp. 121, 123 and 124.

POLES AND ZEROS

Poles and zeros are complex numbers. A pole or a zero can be represented as a point in rectangular coordinates on a map called a *complex plane* or *s-plane*. The abscissa is called the σ-axis (real), and the ordinate is called the $j\omega$-axis (imaginary). A pole on an *s*-plane is denoted by an X, and a zero is denoted by a O. The *s*-plane showing the location of poles and zeroes of $F(s)$ is called a *pole-zero map* of $F(s)$.

The pole–zero map of

$$F(s) = \frac{(s + 2)(s - 3)}{(s + 5)(s + 3 - 2j)(s + 3 + 2j)}$$

is shown below:

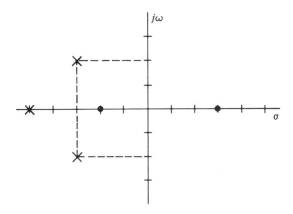

Those values of s for which $F(s) = 0$ are called zeros (numerator), and those values of s for which $F(s) = \infty$ are called poles (denominator).

PARTIAL FRACTIONS

In converting from the s-domain to the time domain, it is sometimes handy to make use of partial fraction expansion. Consider, for example, the following:

$$C(s) = \frac{R(s)\,G(s)}{1 + G(s)\,H(s)}$$

If, by partial fraction expansion, the above equation can be rewritten as:

$$C(s) = \frac{A}{s - P_1} + \frac{B}{s - P_2} + \frac{C}{s - P_3} + \cdots$$

then, converting to the time domain,

$$C(t) = Ae^{P_1 t} + Be^{P_2 t} + Ce^{P_3 t} + \cdots$$

A general algebra book or CRC math tables (31) will show several examples of partial fraction expansion for both nonrepeated and repeated linear factors and also factors of higher degree. In general,

$$\frac{N(X)}{D(X)} = \frac{N(X)}{G(X)\,H(X)\,L(X)} = \frac{A(X)}{G(X)} + \frac{B(X)}{H(X)} + \frac{C(X)}{L(X)}$$

where the numerator $N(X)$ must be of lower order than the denominator $D(X)$. To evaluate the coefficients, multiply both sides of the equation by $D(X)$ to clear fractions. Then collect terms, equate like powers of X, and solve the resulting simultaneous equations for the unknown coefficients.

Example 1. Nonrepeated Linear Factors

$$G(S) = \frac{2s + 1}{s(s + 1)(s + 2)} = \frac{A}{s} + \frac{B}{s + 1} + \frac{C}{s + 2}$$

125

$$2s + 1 = A(s + 1)(s + 2) + Bs(s + 2) + Cs(s + 1)$$

$$A = \frac{2s + 1}{(s + 1)(s + 2)}\bigg|_{s=0} = \frac{1}{2}$$

$$B = \frac{2s + 1}{s(s + 2)}\bigg|_{s=-1} = \frac{-2 + 1}{-1(1)} = 1$$

$$C = \frac{2s + 1}{s(s + 1)}\bigg|_{s=-2} = \frac{-4 + 1}{-2(-1)} = \frac{-3}{2}$$

Example 2.   Repeated Linear Factors

$$G(S) = \frac{s^2 + 2}{s^3(s^2 + 2s - 4)} = \frac{A}{s^3} + \frac{B}{s^2} + \frac{C}{s} + \frac{Ds + E}{s^2 + 2s - 4}$$

A general expression for this type of equation is:

$$\frac{N(X)}{X^m\,G(X)} = \frac{A_0}{X^m} + \frac{A_1}{X^{m-1}} + \cdots + \frac{A_{m-1}}{X} + \frac{F(X)}{G(X)}$$

where

$$F(X) = f_0 + f_1 X + f_2 X^2 + \cdots$$

$$G(X) = g_0 + g_1 X + g_2 X^2 + \cdots$$

$$A_0 = \frac{n_0}{g_0},\ A_1 = \frac{n_1 - A_0 g_1}{g_0},\ A_2 = \frac{n_2 - A_0 g_2 - A_1 g_1}{g_0}$$

$$A_k = \frac{1}{g_0}\left[n_k - \sum_{i=0}^{k-1} A_i g_{k-1}\right]$$

any m: $\quad f_j = n_{m+j} - \sum_{i=0}^{m-1} A_i g_{m+j-i}$

$$N(X) = n_0 + n_1 X + n_2 X^2 + \cdots + n_i X^i$$

Evaluating coefficients of the example,

$$A_0 = A = \frac{n_0}{g_0} = \frac{2}{-4} = -\frac{1}{2},\ A_1 = B = \frac{n_1 - A_0 g_1}{g_0} = \frac{0 + \frac{1}{2}(2)}{-4} = -\frac{1}{4}$$

$$A_2 = C = \frac{n_2 - Ag_2 - Bg_1}{g_0} = \frac{1 - (-\frac{1}{2})(1) - (-\frac{1}{4})(2)}{-4} = -\frac{1}{2}$$

$$m = 3 \begin{cases} D = n_3 - A_0 g_4 - A_1 g_3 - A_2 g_2 = 0 - \left(-\frac{1}{2}\right)(0) - \left(-\frac{1}{4}\right)(0) - \left(-\frac{1}{2}\right)(1) = \frac{1}{2} \\[2mm] E = n_3 - A_0 g_3 - A_1 g_2 - A_2 g_1 = 0 - \left(-\frac{1}{2}\right)(0) - \left(-\frac{1}{4}\right)(1) - \left(-\frac{1}{2}\right)(2) = \frac{5}{4} \end{cases}$$

PROBLEM 5-2. PARTIAL FRACTIONS

Given:

$$C(s) = \frac{8s^2 + 3}{(s^2 + s + 1)(s - 2)} = \frac{As + B}{(s^2 + s + 1)} + \frac{C}{s - 2}$$

$$= \frac{As(s - 2) + B(s - 2) + C(s^2 + s + 1)}{(s^2 + s + 1)(s - 2)}$$

Evaluate A, B, and C.

Solution:

$$8s^2 + 3 = As(s - 2) + B(s - 2) + C(s^2 + s + 1)$$

For $s = 2$, $35 = 7C$, $C = 5$ (answer)

For $s = 0$, $3 = -2B + C = -2B + 5$, $B = 1$ (answer)

For $s = 1$, $11 = -A - B + 3C = -A + 14$, $A = 3$ (answer)

If your answers are correct, go to Problem 5-3.
If your answers are not correct, review p. 125.

PROBLEM 5-3. INVERSE LAPLACE TRANSFORM

Find the inverse Laplace transform of:

$$F(s) = \frac{10\omega^2}{s^2 - \omega^2}$$

Solution:

$$\frac{10\omega^2}{s^2 - \omega^2} = \frac{10\omega^2}{(s + \omega)(s - \omega)} = \frac{A}{(s + \omega)} + \frac{B}{(s - \omega)}$$

$$10\omega^2 = A(s - \omega) + B(s + \omega)$$

For $s = -\omega$, $10\omega^2 = -2\omega A$, $A = -5\omega$
For $s = \omega$, $10\omega^2 = 2B\omega$, $B = 5\omega$

$$F(s) = \frac{-5\omega}{s + \omega} + \frac{5\omega}{s - \omega} = 5\omega \left[\frac{1}{s - \omega} - \frac{1}{s + \omega} \right]$$

$$f(t) = \mathcal{L}^{-1} F(s) = 5\omega(e^{\omega t} - e^{-\omega t}) = 10\omega \sinh \omega t$$ (answer)

The final form is based upon the exponential definition of hyperbolic sine, as found in any book of math tables, such as (31).

If your answer is correct, go to the next section.
If your answers are not correct, review p. 125.

STABILITY

A system is stable if its impulse response approaches zero as time approaches infinity, or if every bounded input produces a bounded output.

For a system to be stable, the real parts of the roots of the characteristic equation must be negative. If they are zero, the system is *marginally stable*, meaning that the impulse response does not decay to zero although it is bounded. Additionally, certain inputs will produce unbounded outputs. Therefore, marginally stable systems are considered to be unstable.

Example:

Let

$$G(s) = K/s, H(s) = H$$

$$\frac{C}{R}(s) = \frac{\dfrac{K}{s}}{1 + \dfrac{HK}{s}} = \frac{K}{s + HK} = K\left[\frac{1}{s + HK}\right]$$

Let $R(s) = 1/s$ (unit step input). Then

$$C(s) = K\left[\frac{1}{s + HK}\right]\left[\frac{1}{s}\right] = \frac{A}{s + HK} + \frac{B}{s} \quad \text{(partial fraction expansion)}$$

$$K = As + B(s + HK)$$

$$B = \frac{K}{s + HK}\bigg|_{s=0} = \frac{1}{H}$$

$$A = \frac{K}{s}\bigg|_{s=-HK} = -\frac{1}{H}$$

$$V_0(s) = \frac{1}{H}\left[\frac{1}{s} - \frac{1}{s + HK}\right]$$

$$V_0(t) = \frac{1}{H}\left[1 - e^{-HKt}\right]$$

This system is stable because as time increases, the exponential term approaches zero.

Stability may be determined by several methods. Some methods indicate only that the system is stable or not stable. Other methods are used to determine *how* stable (or unstable) the system.

Stability may be determined by obtaining the roots of the characteristic equation. Sometimes it is difficult to find the roots of an nth order characteristic equation, in which case graphical methods are normally used. Graphical methods also give a feel for *relative stability*; these methods include:

- Root locus   (time–domain technique)
- Bode Plot
- Nyquist diagram   } (frequency–domain techniques)
- Nichols chart

Nongraphical methods indicate whether or not a system is stable, but not always to what extent. These methods include:

- Routh stability criterion
- Hurwitz stability criterion
- Continued fraction criterion

Graphical methods will be illustrated later.

The Routh stability criterion method of analysis is a simple technique and is defined as follows:

Consider an nth order characteristic equation,

$$a_n s^n + a_{n-1} s^{n-1} + \cdots a_1 s + a_0 = 0$$

ROUTH TABLE

s^n	a_n	a_{n-2}	a_{n-4}	$\cdots$
s^{n-1}	a_{n-1}	a_{n-3}	a_{n-5}	$\cdots$
s^{n-2}	b_1	b_2	b_3	$\cdots$
s^{n-3}	c_1	c_2	c_3	$\cdots$

where $a_n, a_{n-1}, \ldots, a_0$ are coefficients of the characteristic equation and

$$b_1 = \frac{a_{n-1} a_{n-2} - a_n a_{n-3}}{a_{n-1}}$$

$$b_2 = \frac{a_{n-1} a_{n-4} - a_n a_{n-5}}{a_{n-1}}$$

$$c_1 = \frac{b_1 a_{n-3} - a_{n-1} b_2}{b_1}$$

$$c_2 = \frac{b_1 a_{n-5} - a_{n-1} b_3}{b_1}$$

$$b_1 = \frac{c_1 b_2 - b_1 c_2}{c_1}$$

All roots of this characteristic equation have negative real parts if, and only if, the elements of the first column have the same sign. Otherwise, the number of roots with positive real parts is equal to the number of sign changes. If a zero occurs in the first column, the system is also unstable.

PROBLEM 5-4. STABILITY, ROUTH

Using the Routh stability criterion, determine the range of values of K for which the control system shown below is stable.

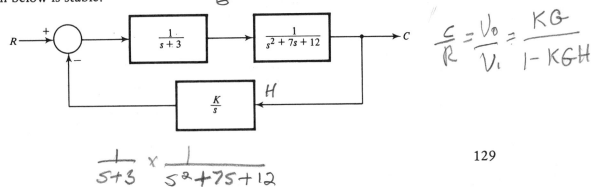

$$\frac{C}{R} = \frac{V_0}{V_i} = \frac{KG}{1 - KGH}$$

$$\frac{1}{s+3} \times \frac{1}{s^2 + 7s + 12}$$

129

Solution:

$$GH = \frac{K}{s(s + 3)(s^2 + 7s + 12)}$$

The characteristic equation is:

$$D_{GH} \pm N_{GH} = 0$$

$$s^4 + 10s^3 + 33s^2 + 36s + K = 0$$

$$s^3 + 7s^2 + 12s$$
$$3s^2 + 21s + 36$$
$$\overline{s^3 + 10s^2 + 33s + 36}$$

s^4	1	33	K
s^3	10	36	0
s^2	29.4	K	0
s^1	36 − 0.34K	0	0
s^0	K		

$a_n = 1$

$a_{n-1} = 10$

$a_{n-2} = 33$

$a_{n-3} = 36$

$a_{n-4} = K$

$$a_n = 1, \quad a_{n-1} = 10, \quad a_{n-2} = 33, \quad a_{n-3} = 36, \quad a_{n-4} = K$$

$$b_1 = \frac{a_{n-1}a_{n-2} - a_n a_{n-3}}{a_{n-1}} = \frac{330 - 36}{10} = \frac{294}{10} = 29.4$$

$$b_2 = \frac{a_{n-1}a_{n-4} - a_n a_{n-5}}{a_{n-1}} = \frac{10K - 1 \times 0}{10} = K$$

$$c_1 = \frac{b_1 a_{n-3} - a_{n-1}b_2}{b_1} = \frac{29.4 \times 36 - 10K}{29.4} = 36 - \frac{10}{29.4}K$$

All signs in the first column must be positive for stability. Therefore:

$$K > 0 \quad \text{and} \quad 36 - 0.34K > 0$$

$$K < \frac{36}{0.34} = 105.88$$

(answer)   For stability, $0 < K < 105.88$.

If your answer is correct go to Problem 5-5.
If your answer is not correct, review p. 129.

PROBLEM 5-5. ERROR AND STABILITY

For the control system that follows, determine the steady-state error for a step input and the range of K for which this system is stable.

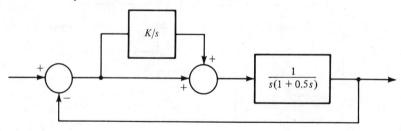

Solution:

Using block diagram algebra, simplify the preceding diagram:

$$1 + \frac{K}{s} = \frac{s + K}{s}$$

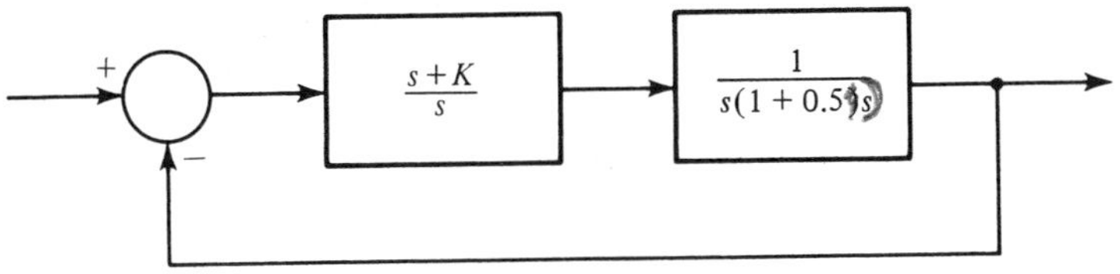

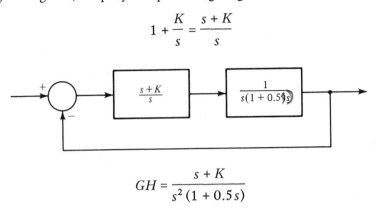

$$GH = \frac{s + K}{s^2 (1 + 0.5s)}$$

Now determine the system type. DiStefano[3] has a good, brief description of system type. In general, for type ℓ, $\ell = b - a$ for $b \geqslant a$,

where

$$GH = \frac{Ks^a \prod_{i=1}^{m-a} (s + Z_i)}{s^b \prod_{i=1}^{n-b} (s + P_i)} = \frac{KB_1(s)}{s^\ell B_2(s)}$$

In this problem, $a = 0$ and $b = 2$; therefore, $\ell = b - a = 2 - 0 = 2$, so this is a type 2 system. As shown in the table on p. 165 of DiStefano[3], the steady-state error for a type 2 system is zero. (answer)

The characteristic equation is:

$$D_{GH} \pm N_{GH} = s^2 (1 + 0.5s) + s + K = 0$$

$$s^3 + 2s^2 + 2s + 2K = 0$$

Routh Table

s^3	1	2
s^2	2	$2K$
s^1	$2 - K$	0
s^0	$2K$	

To assure the sign for each term in Column 1 is the same (positive) for stability:

$$2 - K > 0, K < 2$$

also,

$$2K > 0, K > 0$$

Thus, the range on K_1 is

$$0 < K < 2$$ (answer)

If your answers are correct, go on to the next section.
If your answers are not correct, study DiStefano[3] and pp. 128 and 129.

TRANSFER FUNCTION

The transfer function of a system is that factor $P(s)$ in the equation for $Y(s)$ multiplying the transform of the input $X(s)$.

$$Y(s) = P(s)\ X(s) + \underbrace{\text{(terms due to initial conditions)}}$$

$$= 0 \text{ if system is at rest prior}$$
$$\text{to application of the input}$$

This equation separates the forced response excluding initial values (on the left) from the free response and the forced response initial value terms (on the right). If all initial values are zero, the equation becomes:

$$Y(s) = P(s)\ X(s)$$

and

$$Y(t) = \mathcal{L}^{-1}\ [P(s)\ X(s)]$$

$$P(s) = \frac{Y(s)}{X(s)}$$

Not all transfer functions are rational algebraic expressions. The transfer function of a system having time delays contains terms of the form e^{-sT}. The transfer function of pure time delay is $P(s) = e^{-sT}$, where T is the time delay in units of time.

PROBLEM 5-6. TRANSFER FUNCTION

A servomechanism for controlling angular position by means of differentially connected potentiometers consists of a servoamplifier, a servomotor, and a feedback potentiometer whose transfer functions are as follows:

$$G_a = \frac{1}{s + 3}, \quad G_m = \frac{1}{s + 2}$$

$$H = K$$

Determine:

   a. open loop transfer function;
   b. system transfer function; and
   c. range of K for a stable system.

Solution:

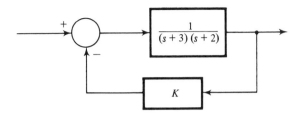

a. Open loop transfer function = GH

$$= \frac{K}{(s + 3)(s + 2)}$$ (answer)

b. System transfer function = $\dfrac{G}{1 + GH}$

$$= \frac{\dfrac{1}{(s + 3)(s + 2)}}{1 + \dfrac{K}{(s + 3)(s + 2)}} = \frac{1}{K + (s + 3)(s + 2)}$$

$$= \frac{1}{s^2 + 5s + (6 + K)}$$ (answer)

c. For the system to be stable there must be no negative factors in the denominator. Therefore, $(6 + K)$ must always be positive. Hence,

$$(6 + K) > 0$$
$$K > -6$$ (answer)

If your answers are correct, go on to the next section.
If your answers are not correct, review p. 121.

COMPENSATION

Compensation networks are sometimes introduced into a control system to improve its operation. Compensators may be introduced into either the forward path or the feedback path in order to achieve the desired results. Compensation may be either passive or active.

Three commonly used control system compensators using passive RC networks are illustrated in the following paragraphs.

Lead Compensator

The transfer function of a lead compensator is:

$$P_{lead}(s) = \frac{s + a}{s + b}$$

zero $= -a = -\left[\dfrac{1}{R_1 C}\right]$

pole $= -b = -\left[\dfrac{1}{R_1 C} + \dfrac{1}{R_2 C}\right]$

$b > a$

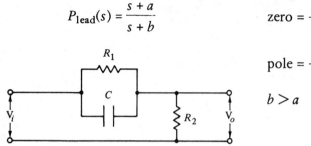

Lag Compensator

The transfer function of a lag compensator is:

$$P_{lag}(s) = \frac{a\,(s + b)}{b\,(s + a)}$$

zero at $s = -b = -\left[\dfrac{1}{R_2 C}\right]$

pole at $s = -a = -\left[\dfrac{1}{(R_1 + R_2)C}\right]$

gain factor $= \dfrac{a}{b}$

$b > a$

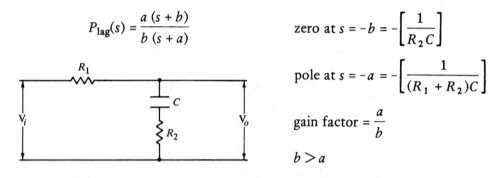

Lag–Lead Compensator

The transfer function of a lag–lead compensator is:

$$P_{ll}(s) = \frac{(s + a_1)\,(s + b_2)}{(s + b_1)\,(s + a_2)}$$

zeros at $-a_1, -b_2$

poles at $-b_1, -a_2$

$a_1 b_2 = b_1 a_2$

$b_1 > a_1, b_2 > a_2$

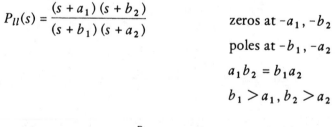

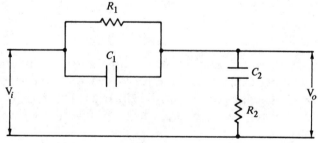

Simple Lag Compensator

Shown below is a simple lag compensator with its transfer function:

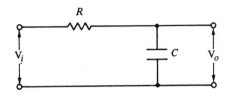

$$P(s) = \frac{V_0(s)}{V_i(s)} = \frac{\dfrac{1}{Cs}}{R + \dfrac{1}{Cs}} = \frac{\dfrac{1}{RC}}{s + \dfrac{1}{RC}}$$

Example. System Compensation

Consider the unstable second order system shown below:

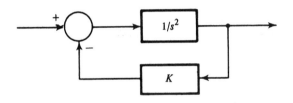

$$\frac{C}{R}(s) = \frac{1/s^2}{1 + K/s^2} = \frac{1}{s^2 + K}, \quad s = \pm j\sqrt{K}$$

This system is characterized by the following:

- crosses 0 db at -40 db/decade
- phase margin $= 0°$
- poles on $j\omega$ axis
- system oscillates

Now, add lead compensation to the forward path as shown below:

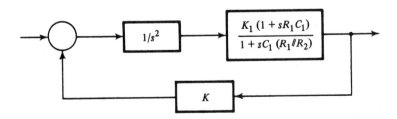

Or add compensation in the feedback loop as shown below:

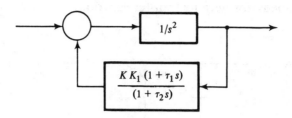

The system is stable in either configuration.

PROBLEM 5-7. CANCELLATION COMPENSATION

Using cancellation compensation in the feedback control system that follows, design a lead network that will yield a system having a damping ratio, ζ, of 0.7 and an undamped natural frequency, ω_n, of 4. Calculate the resulting system gain.

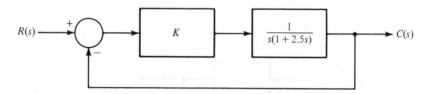

Solution:

$$\frac{C_{(s)}}{R(s)} = \frac{\dfrac{0.4K}{s(s + 0.4)}}{1 + \dfrac{0.4K}{s(s + 0.4)}} = \frac{0.4K}{s^2 + 0.4s + 0.4K}$$

This is of the form:

$$\frac{0.4K}{s^2 + 2\zeta\omega_n s + \omega_n^2}$$

Thus,

$$\omega_n = \sqrt{0.4K} = 4, \ K = 40$$

$2\zeta\omega_n s = 0.4s, \ \zeta = \dfrac{0.4}{2\omega_n} = 0.01$, which does not satisfy the requirement that $\zeta = 0.7$.

Therefore, add lead compensation to the forward path as follows:

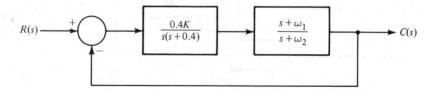

For cancellation, $(s + \omega_1)$ cancels out $(s + 0.4)$, $\omega_1 = 0.4$, $\dfrac{s + \omega_1}{s + 0.4} = 1$

$$\frac{C(s)}{R(s)} = \frac{\dfrac{(0.4K)\,(s + \omega_1)}{s(s + 0.4)\,(s + \omega_2)}}{1 + \dfrac{(0.4K)\,(s + \omega_1)}{\omega_s(s + 0.4)\,(s + \omega_2)}} = \frac{0.4K(s + \omega_1)}{s(s + 0.4)\,(s + \omega_2) + 0.4K\,(s + \omega_1)}$$

$$= \frac{0.4K}{s(s + \omega_2) + 0.4K} = \frac{0.4K}{s^2 + \omega_2 s + 0.4K}$$

$$= \frac{0.4K}{s^2 + 2\zeta\omega_n s + \omega_n^2}$$

$$\zeta = 0.7 = \frac{\omega_2}{2\omega_n}, \quad \omega_2 = 0.7 \times 2 \times 4 = 5.6 \text{ rad/sec}$$

Forward path lead compensation $= G_C(s) = \dfrac{s + 0.4}{s + \cancel{0.7}\,5.6}$

For the compensated system:

$$G(s) = \left(\frac{s + 0.4}{s + 5.6}\right)\,(16)\,\left(\frac{1}{s(s + 0.4)}\right) = \frac{16}{s(s + 5.6)} = \frac{16/5.6}{s(1 + s/5.6)} = 2.86$$

System gain = 2.86.   (answer)

If your answer is correct, go on to problem 5-8.
If your answer is not correct, review pp. 133–136.

PROBLEM 5-8.   LEAD COMPENSATION

The performance of an automatic control system is to be improved by use of a phase lead compensator. If the input to the network that follows is of the form $e_1 = E_m \sin wt$ and the desired output is of the form $e_2 = KE_m \sin(wt + \theta)$, determine the general relationship for the phase lead angle, θ, and the amplifier gain $1/K$ in terms of R_1, R_2, and R_3. Assume the amplifier has an infinite input impedance.

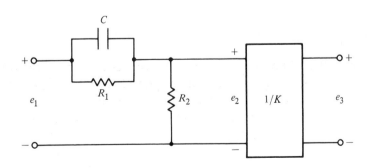

Solution:

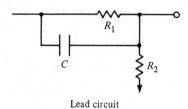

Lead circuit

$e_1 = E_m \sin wt$

$e_2 = KE_m \sin (wt + \theta)$

$e_3 = E_m \sin (wt + \theta)$

Transfer function $= K \dfrac{(T_1 s + 1)}{(T_2 s + 1)}$

First break $= T_1 = R_1 C$

Second break $= T_2 = \dfrac{R_1 R_2 C}{R_1 + R_2}$

$$K = \frac{R_2}{R_1 + R_2} \quad \text{(voltage divider)}$$

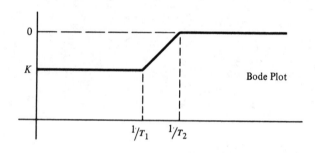

Bode Plot

(answer)

$$\sin \theta = \frac{|K - 1|}{|K + 1|}$$

If your answer is correct, go on to the next section.
If your answer is not correct, review pp. 133–140.

BODE ANALYSIS

Bode analysis is a graphical technique which can be used to determine the relative stability of a system. Bode plots consist of two graphs: the magnitude of $GH(j\omega)$, and the phase angle of $GH(j\omega)$, both plotted as a function of frequency, $j\omega$. Log scales are usually used for the frequency axes and for $|GH(j\omega)|$. Gain and phase margins are often defined in terms of Bode plots.

Example:

Given:

$$G(s) = \frac{K_1}{(s + \omega_1)(s + \omega_2)}$$

$$H(s) = \frac{K_2(s + \omega_3)}{(s + \omega_4)(s + \omega_5)}$$

$$GH(s) = \frac{\dfrac{K_1 K_2 \omega_3 \left(1 + \dfrac{s}{\omega_3}\right)}{\omega_1 \omega_2 \omega_4 \omega_5}}{\left(1 + \dfrac{s}{\omega_1}\right)\left(1 + \dfrac{s}{\omega_2}\right)\left(1 + \dfrac{s}{\omega_4}\right)\left(1 + \dfrac{s}{\omega_5}\right)}$$

$$GH(j\omega) = \frac{K_3\left(1 + j\dfrac{\omega}{\omega_3}\right)}{\left(1 + j\dfrac{\omega}{\omega_1}\right)\left(1 + j\dfrac{\omega}{\omega_2}\right)\left(1 + j\dfrac{\omega}{\omega_4}\right)\left(1 + j\dfrac{\omega}{\omega_5}\right)}$$

where

$$K_3 = \frac{K_1 K_2 \omega_3}{\omega_1 \omega_2 \omega_4 \omega_5}$$

ϕ = phase shift

$$= -\arctan\frac{\omega}{\omega_1} - \arctan\frac{\omega}{\omega_2} + \arctan\frac{\omega}{\omega_3} - \arctan\frac{\omega}{\omega_4} - \arctan\frac{\omega}{\omega_5}$$

plot $20 \log |GH(j\omega)|$ vs $\log \omega$.
  For $\omega \ll \omega_i$,

$$\text{gain of } \left(1 + j\frac{\omega}{\omega_i}\right) = 1$$

For $\omega \gg \omega_i$,

$$\text{gain of } \left(1 + j\frac{\omega}{\omega_i}\right) = \frac{\omega}{\omega_i}$$

$$\text{gain of } \frac{1}{1 + j\dfrac{\omega}{\omega_i}} = -20 \text{ db/decade}$$

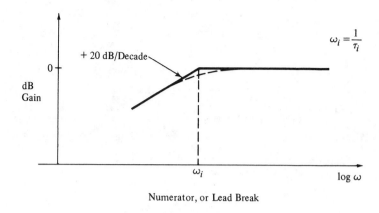

$$\omega_i = \frac{1}{\tau_i}$$

Numerator, or Lead Break

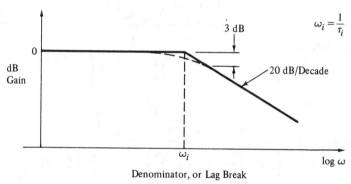

$$\omega_i = \frac{1}{\tau_i}$$

Denominator, or Lag Break

For the composite Bode plot,

$$\text{gain total} = \sum \text{individual gain plots}$$

$$= \log G_1 + \log G_2 - \log G_3 - \log G_4 - \log G_5$$

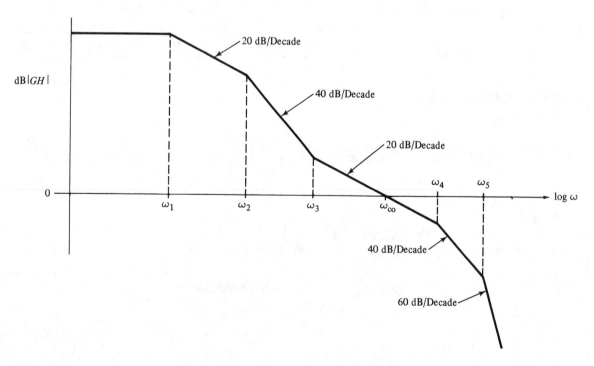

140

The system is stable if the Bode plot is characterized as follows:

- Slope of gain plot is -20 db/decade at ω_{co} (db$|GH|$ = 0 db)
- Total phase lag $< 180°$ at ω_{co}

Thus, 1st order has $-90°$ and is stable, while 2nd order has $-90° > \phi > -180°$.

PROBLEM 5-9. BODE ANALYSIS

A control system has the following block diagram:

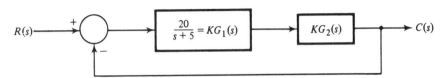

The frequency response of $KG_2(s)$ has been measured and plotted on the graph below:

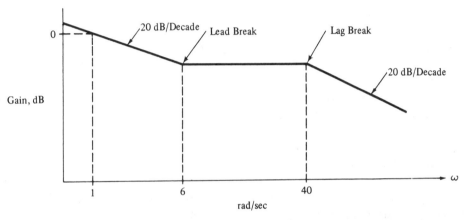

Determine the system transfer function $\dfrac{C}{R}(s)$.

Solution:

From the Bode plot it is seen that there is a lead break at $\omega = 6$ and a lag break at $\omega = 40$. Therefore, the transfer function for $KG(s)$ is:

$$KG_2(s) = \frac{K(s+6)}{s(s+40)} = \frac{K(1+0.167s)\,6}{s(1+0.025s)\,40} \longrightarrow \frac{0.15K(1+j0.167\omega)}{j\omega(1+j0.025\omega)}$$

$$= \frac{0.15K(1+j0.167\omega)}{-0.025\omega^2 + j\omega}$$

To find K, evaluate gain = 0 db at $\omega = 1$.

$$|KG_2(s)| = \frac{0.15K(1+j0.167)}{(-0.025+j1)} = \frac{0.15K(1.01)}{1} = 0.15K$$

$$\text{db} = 20 \log 0.15K \equiv 0, \quad 0.15K = 1, \quad \therefore\ K = 6.67$$

141

Thus,

$$KG_2(s) = \frac{6.67(s + 6)}{s(s + 40)}$$

Evaluating the system transfer function,

$$\frac{C}{R}(s) = \frac{KG}{1 + KGH} = \frac{KG}{1 + KG} \text{ for } H = 1 \quad = \quad \frac{1}{1 + \frac{1}{KG}}$$

$$\frac{C}{R}(s) = \frac{1}{1 + \dfrac{s(s + 40)(s + 5)}{6.67(s + 6)\,20}} = \frac{133(s + 6)}{133(s + 6) + s(s + 40)(s + 5)}$$

(answer)
$$= \frac{133(s + 6)}{s^3 + 45s^2 + 333s + 800}$$

$$KG = \frac{20}{(S+5)} \times \frac{6.67(S+6)}{S(S+40)}$$

$$\frac{1}{K2} = \frac{(S+5)(S+40)S}{20 \times 6.67 (S+6))}$$

If your answer is correct go to the next section.
If your answer is not correct, review p. 139.

ROOT LOCUS

Root locus analysis of a control system is an analytical method for displaying the location of the poles and zeros of the closed-loop transfer function as a function of the gain factor K of the open loop transfer function. Root locus analysis yields accurate time-domain response characteristics while also providing frequency response information.

PROBLEM 5-10. ROOT LOCUS

Given the feedback control system defined by the block diagram shown below, determine using root locus analysis whether or not the system is stable.

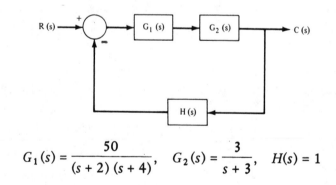

$$G_1(s) = \frac{50}{(s + 2)(s + 4)}, \quad G_2(s) = \frac{3}{s + 3}, \quad H(s) = 1$$

Solution:

The open loop transfer function is:

$$GH = \frac{K}{(s + 2)(s + 4)(s + 3)}$$

where $K = 150$.

The characteristic equation is,

$$D_{GH} \pm N_{GH} = 0$$

$$150 + (s + 2)(s + 4)(s + 3) = 0$$

$$150 + s^3 + 9s^2 + 26s + 24 = 0$$

$$s^3 + 9s^2 + 26s + 174 = 0$$

From the open loop transfer function, open loop poles (there are no zeros) are located as shown on the s-plane below:

Calculating the breakaway point, σ_b :

$$\frac{1}{(\sigma_b + 2)} + \frac{1}{(\sigma_b + 4)} + \frac{1}{(\sigma_b + 3)} = 0$$

$$(\sigma_b + 4)(\sigma_b + 3) + (\sigma_b + 2)(\sigma_b + 3) + (\sigma_b + 2)(\sigma_b + 4) = 0$$

$$3\sigma_b^2 + 18\sigma_b + 26 = 0$$

$$\sigma_b^2 + 6\sigma_b + \frac{26}{3} = 0$$

Using the quadratic formula*,

$$\sigma_b = \frac{-6 \pm \sqrt{36 - \dfrac{4 \times 26}{3}}}{2} = -3 \pm 0.58 = -3.58 \quad \text{or} \quad -2.43$$

Select -2.43 since it lies between the poles at -2 and -3.

Calculating the center of the asymptotes:

$$\sigma_c = -\frac{\sum\limits_{i=1}^{n} P_i - \sum\limits_{i=1}^{m} Z_i}{n - m}$$

where:

n = number of poles = 3
m = number of zeros = 0

$$\therefore \; \sigma_c = +\frac{9 - 0}{3} = +3$$

Calculating the angle between asymptotes and the real axis:

$$\beta = \frac{(2N + 1)\,180}{n - m} \; \text{degrees} \quad \text{for} \quad K > 0$$

$$* \; x = \frac{-b \pm \sqrt{b^2 - 4ac}}{2a}$$

$$4 - 1 = 3$$

$$\beta = \frac{(2N+1)\,180}{3} \qquad N = 1$$

143

where N is any arbitrary integer (in this case let $N = 0$)

$$\beta = \frac{180°}{3} = 60°$$

Determining the value of K at the crossing of root locus on the imaginary axis:
For the characteristic equation,

$$s^3 + 9s^2 + 26s + 24 + K = 0$$

let $s = j\omega$ and rewrite,

$$-j\omega^3 - 9\omega^2 + 26j\omega + 24 + K = 0$$

Separating the real and imaginary parts and equating to zero,

$$-9\omega^2 + 24 + K = 0, \quad K = 9\omega^2 - 24$$

$$-\omega^3 + 26\omega = 0, \quad \omega^2 = 26$$

Thus, the root locus crosses the imaginary axis at $\omega = \sqrt{26} = 5.1$, and K at this point is:

$$K = 9 \times 26 - 24 = 210$$

(answer)   For stability K must be less than 210 in the LHP. Since $K = 150$, the system is stable.
The figure below is the final plot of the root loci based on the preceding calculations.

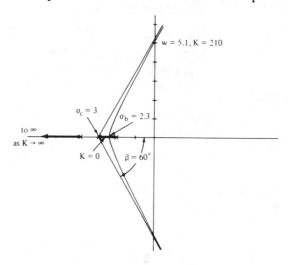

Alternate Solution:

The system may be checked for stability using alternative methods. One quick method is by use of the Routh criteria shown below:

$$s^3 + 9s^2 + 26s + 174 = 0$$

ROUTH TABLE

s^3	1	26	0
s^2	9	174	
s^1	6.67	0	
s^0	174		

144

$$b_1 = \frac{(9)(26) - (1)(174)}{9} = 6.67$$

$$b_2 = \frac{(9)(0) - (1)(0)}{9} = 0$$

$$C_1 = \frac{(6.67)(174) - (9)(0)}{6.67} = 174 \qquad \text{(answer)}$$

Since there are no sign changes in the first column, the system is stable.

If your answer is correct, go to the next chapter.
If your answer is not correct, review pp. 129, 130, 143 and 144.

6 Electronics

Discussion of and problems related to diode suppression, one-stage transistor amplifier, common base amplifier, two-stage transistor amplifier, transistor curves and load line, transistor stability, field effect transistor amplifier, vacuum tube amplifier, operational amplifer, and amplifier class.

INTRODUCTION

Only the most basic aspects of electronics are presented as they relate to the type of problems found in the PE Exam. Reference to a basic college text on electronics may be helpful in reviewing this subject.

BLACK BOX ANALYSIS

A circuit may be represented by a black box having an input and an output with its parameters expressed in several ways, as illustrated by the models below.

General Two-Port (Four-Terminal) Network

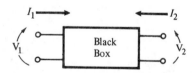

Equations	Parameter	Units
$V = ZI$	Impedance	ohms (Ω)
$I = YV$	Admittance	mhos ($\mho$)
$V_1 = f(I_1, V_2)$ $I_2 = f(I_1, V_2)$	Hybrid	$\Omega,\ \mho$ V/V, A/A

Open-Circuit Impedance Parameters

$$\begin{bmatrix} V_1 \\ V_2 \end{bmatrix} = \begin{bmatrix} Z_{11} & Z_{12} \\ Z_{21} & Z_{22} \end{bmatrix} \begin{bmatrix} I_1 \\ I_2 \end{bmatrix} \longrightarrow \begin{array}{l} V_1 = Z_{11}I_1 + Z_{12}I_2 \\ V_2 = Z_{21}I_1 + Z_{22}I_2 \end{array}$$

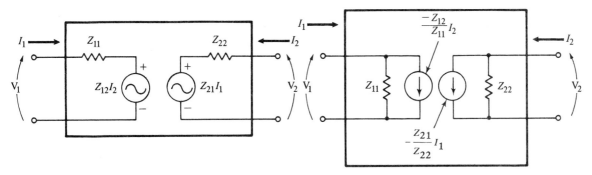

Short-Circuit Admittance Parameters

$$\begin{bmatrix} I_1 \\ I_2 \end{bmatrix} = \begin{bmatrix} Y_{11} & Y_{12} \\ Y_{21} & Y_{22} \end{bmatrix} \begin{bmatrix} V_1 \\ V_2 \end{bmatrix} \longrightarrow \begin{array}{l} I_1 = Y_{11}V_1 + Y_{12}V_2 \\ I_2 = Y_{21}V_1 + Y_{22}V_2 \end{array}$$

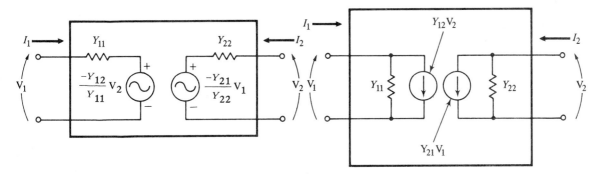

Hybrid Parameters

$$\begin{bmatrix} V_1 \\ I_2 \end{bmatrix} = \begin{bmatrix} b_{11} & b_{12} \\ b_{21} & b_{22} \end{bmatrix} \begin{bmatrix} I_1 \\ V_2 \end{bmatrix} \qquad \begin{array}{l} V_1 = b_{11}I_1 + b_{12}V_2 \\ I_2 = b_{21}I_1 + b_{22}V_2 \end{array}$$

Parameter	Units
b_{11}	Ω
b_{12}	V/V
b_{21}	A/A
b_{22}	$\mho$

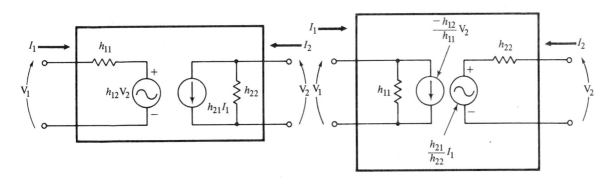

DIODES

The classical diode equation is:

$$I = I_s(e^{qV/kT} - 1)$$

where

I_s = saturation current
q = charge on an electron = 1.602 × 10^{-19} coulomb
k = Boltzmann's constant = 1.38 × 10^{-16} erg/°K
T = temperature in °K (°K = °C + 273)
V = forward voltage across the diode

If V is about four times greater than kT/q (say 0.1 volt), the diode equation simplifies to:

$$I = I_s(e^{qV/kT})$$

and the dynamic forward resistance at a specific operating point is:

$$R_f = \frac{kT}{qI} \ \Omega$$

At room temperature, kT/q is approximately 26 mV; thus, a diode with 1 mA of forward current will exhibit 26 ohms of dynamic resistance.

The following are diode models, characteristics, and applications.

Diode Symbol and Characteristics

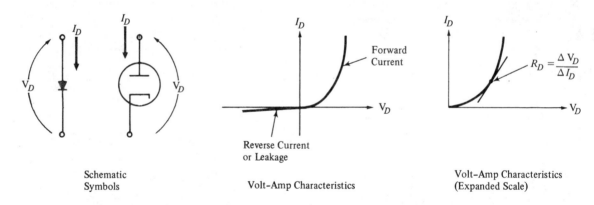

Schematic Symbols

Volt-Amp Characteristics

Volt-Amp Characteristics (Expanded Scale)

Diode Circuit Model

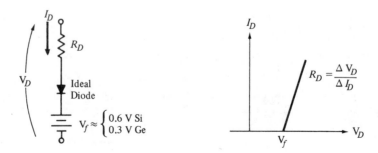

Zener Diode

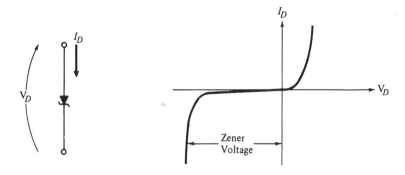

Zener Model

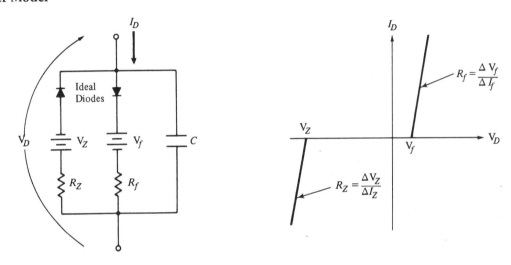

PROBLEM 6-1. DIODE SUPPRESSION

A relay coil is to be protected against excessively high induced voltage when it is open-circuited. The circuit is shown below. Show the correct connection of the diode and specify its minimum voltage and current ratings.

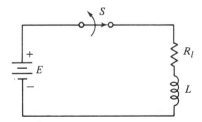

Solution:

The diode is connected as shown below so that it will conduct when the switch is open, thus permitting the coil current to decay to zero while preventing arcing of the switch contacts.

149

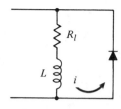

(answer) Maximum voltage across the diode is E volts, and maximum current through the diode is E/R amps. To provide a conservative safety margin, a diode should be selected that can withstand twice these values.

If your answer is correct, continue reading.
If your answer is not correct, review p. 148.

TRANSISTORS

General

Bipolar transistors may be used in three configurations: common emitter, common base, and common collector (emitter-follower). Transistor models with their parameters expressed in different ways are delineated below. Several aspects of transistors must be considered in practical designs, including temperature variation, leakage current, gain–bandwidth, maximum operating frequency, operating load current, biasing, and stability.

Temperature variation can affect current gain, collector leakage, and power dissipation. As a rule of thumb, collector leakage I_{co} doubles with every Δt temperature increase, where $\Delta t = 10°C$ for germanium and $6°C$ for silicon. It also increases with increased collector voltage. The *gain–bandwidth product*, f_T, is a common emitter parameter; it is the frequency at which gain b_{fe}, drops to unity (0 db). Although common emitter current gain is 0 db at f_T, there may still be considerable power gain at f_T due to different input and output impedance levels. Thus, f_T is not necessarily the maximum useful operating frequency. In a practical circuit, the minimum load current should be no less than 10 times the leakage current. The maximum load current should not exceed the maximum power dissipation voltage. The operating load current should be midway between these two extremes. When the operating load current is flowing at the selected no-signal point, the collector load resistance should drop the collector voltage to one-half the supply voltage, for class A operation.

Transistor Equivalent Circuits

Transistor equivalent circuits in the common emitter, common base, and common collector configurations are shown below. Direction of current flow depends upon whether the transistor is PNP or NPN. Table 6-1 delineates the small signal characteristics; it provides for conversion between T-parameters and b-parameters in either direction for the three circuit configurations, and it gives typical actual parameter values. Table 6-2 summarizes the b-parameters for the three circuit configurations.

150

TABLE 6-1. Transistor Small Signal Characteristics.

(Numerical values are for a typical transistor operating under standard conditions)

Symbols			Common Emitter	Common Base	Common Collector	T-Equivalent
h_{11e}	h_{ie}	r_i	1500 ohms	$\dfrac{h_{ib}}{1+h_{fb}}$	h_{ic}	$r_b + \dfrac{r_e}{1-\alpha}$
h_{12e}	h_{re}		3×10^{-4}	$\dfrac{h_{ib}h_{ob}}{1+h_{fb}} - h_{rb}$	$1 - h_{rc}$	$\dfrac{r_e}{(1-\alpha)\,r_c}$
h_{21e}	h_{fe}	A_I	49	$-\dfrac{h_{fb}}{1+h_{fb}}$	$-(1+h_{fc})$	$\dfrac{\alpha}{1-\alpha} = \beta$
h_{22e}	h_{oe}	$\dfrac{1}{r_o}$	30×10^{-6} mho	$\dfrac{h_{ob}}{1+h_{fb}}$	h_{oc}	$\dfrac{1}{(1-\alpha)\,r_c} = \dfrac{1}{r_d}$
h_{11b}	h_{ib}	r_i	$\dfrac{h_{ie}}{1+h_{fe}}$	30 ohms	$-\dfrac{h_{ic}}{h_{fc}}$	$r_e + r_b(1-\alpha)$
h_{12b}	h_{rb}		$\dfrac{h_{ie}h_{oe}}{1+h_{fe}} - h_{re}$	5×10^{-4}	$h_{re} - 1 - \dfrac{h_{ic}h_{oc}}{h_{fc}}$	$\dfrac{r_b}{r_c}$
h_{21b}	h_{fb}	A_I	$-\dfrac{h_{fe}}{1+h_{fe}}$	-0.98	$-\dfrac{1+h_{fc}}{h_{fc}}$	$-\alpha$
h_{22b}	h_{ob}	$\dfrac{1}{r_o}$	$\dfrac{h_{oe}}{1+h_{fe}}$	0.5×10^{-6} mho	$\dfrac{h_{oc}}{h_{fc}}$	$\dfrac{1}{r_c}$
h_{11c}	h_{ic}	r_i	h_{ie}	$\dfrac{h_{ib}}{1+h_{fb}}$	1500 ohms	$r_b + \dfrac{r_e + R_l}{1-\alpha} = R_l(\beta + 1)$
h_{12c}	h_{rc}		$1 - h_{re} \approx 1$	1	1	$1 - \dfrac{r_e}{(1-\alpha)\,r_c}$
h_{21c}	h_{fc}	A_I	$-(1+h_{fe})$	$-\dfrac{1}{1+h_{fb}}$	-50	$-\dfrac{1}{1-\alpha} = -(\beta + 1)$
h_{22c}	h_{oc}	$\dfrac{1}{r_o}$	h_{oe}	$\dfrac{h_{ob}}{1+h_{fb}}$	30×10^{-6} mho	$\dfrac{1}{(1-\alpha)\,r_c}$
α			$\dfrac{h_{fe}}{1+h_{fe}}$	$-h_{fb} = -h_{21b}$	$\dfrac{1+h_{fc}}{h_{fc}}$	0.98
r_c			$\dfrac{1+h_{fe}}{h_{oe}}$	$\dfrac{1-h_{rb}}{h_{ob}} = \dfrac{1}{h_{22b}}$	$-\dfrac{h_{fc}}{h_{oc}}$	1.7 MΩ
r_e			$\dfrac{h_{re}}{h_{oe}}$	$h_{ib} - \dfrac{h_{rb}}{h_{ob}}(1+h_{fb})$	$\dfrac{1-h_{rc}}{h_{oc}}$	10 Ω
r_b			$h_{ie} - \dfrac{h_{re}}{h_{oe}}(1+h_{fe})$	$\dfrac{h_{rb}}{h_{ob}} = \dfrac{h_{12b}}{h_{22b}}$	$h_{ic} + \dfrac{h_{fc}}{h_{oc}}(1-h_{rc})$	1 KΩ

r_i = input resistance r_c = collector resistance   power gain = $G = A_V A_I$

r_o = output resistance r_e = emitter resistance $\beta = \dfrac{\alpha}{1-\alpha}, \quad \alpha = \dfrac{\beta}{\beta+1}$

A_V = voltage amplification r_b = base resistance R_l = load resistance

A_I = current amplification α = short circuit current multiplier

TABLE 6-2. h-Parameters.

Symbol	Definition	General	Common Emitter	Common Base	Common Collector
h_{11}	Z_{in} with output shorted	$\dfrac{\Delta V_1}{\Delta I_1}\Big|_{\Delta V_2=0}$	$\dfrac{\Delta V_{BE}}{\Delta I_B}\Big|_{\Delta V_{CE}=0} = \dfrac{h_{11}}{(1+h_{21})}$	$\dfrac{\Delta V_{EB}}{\Delta I_E}\Big|_{\Delta V_{CB}=0} = h_{11}$	$\dfrac{\Delta V_{CB}}{\Delta I_B}\Big|_{\Delta V_{CB}=0} = \dfrac{h_{11}}{(1+h_{21})}$
h_{12}	inverse voltage transfer ratio with input open	$\dfrac{\Delta V_1}{\Delta V_2}\Big|_{\Delta I_1=0}$	$\dfrac{\Delta V_{BE}}{\Delta V_{CE}}\Big|_{\Delta I_B=0} = \dfrac{(\Delta h_e - h_{12})}{(1+h_{21})}$	$\dfrac{\Delta V_{EB}}{\Delta V_{CB}}\Big|_{\Delta I_E=0} = h_{12}$	$\dfrac{\Delta V_{CB}}{\Delta V_{EC}}\Big|_{\Delta I_B=0} = 1$
h_{21}	forward current transfer ratio with output shorted	$\dfrac{\Delta I_2}{\Delta I_1}\Big|_{\Delta V_2=0}$	$\dfrac{\Delta I_C}{\Delta I_B}\Big|_{\Delta V_{CE}=0} = -\dfrac{h_{21}}{(1+h_{21})}$	$\dfrac{\Delta I_C}{\Delta I_E}\Big|_{\Delta V_{CB}=0} = h_{21}$	$\dfrac{\Delta I_E}{\Delta I_B}\Big|_{\Delta V_{CB}=0} = -\dfrac{1}{(1+h_{21})}$
h_{22}	Y_{out} with input open	$\dfrac{\Delta I_2}{\Delta V_2}\Big|_{\Delta I_1=0}$	$\dfrac{\Delta I_C}{\Delta V_{CE}}\Big|_{\Delta I_B=0} = \dfrac{h_{22}}{(1+h_{21})}$	$\dfrac{\Delta I_C}{\Delta V_{CB}}\Big|_{\Delta I_E=0} = h_{22}$	$\dfrac{\Delta I_E}{\Delta V_{EC}}\Big|_{\Delta I_B=0} = \dfrac{h_{22}}{(1+h_{21})}$
A_V	e_o/e_i	$\dfrac{R_l}{h_{11}+\Delta h R_l}$	$\dfrac{-h_{fe}}{h_{ie}G_l+\Delta h_e}$	$\dfrac{-h_{fb}}{h_{ib}G_l+\Delta h_b}$	$\dfrac{-h_{fc}}{h_{ic}G_l+\Delta h_c}$
R_i	input resistance	$\dfrac{h_{11}+\Delta h R_l}{1+h_{22}R_l}$	$h_{ie} - \dfrac{h_{fe}h_{re}}{h_{oe}+G_l}$	$h_{ib} - \dfrac{h_{fb}h_{rb}}{h_{ob}+G_l}$	$h_{ic} - \dfrac{h_{fc}h_{rc}}{h_{oc}+G_l}$
R_o	output resistance	$\dfrac{h_{11}+R_g}{\Delta h + h_{22}R_g}$	$\dfrac{R_g+h_{ie}}{h_{oe}(R_g+h_{ie})-h_{fe}h_{re}}$	$\dfrac{h_{ib}+R_g+R_B(1+h_{fb})}{\Delta h_b'+R_g h_{ob}}$	$\dfrac{R_g+h_{ic}}{R_g h_{oc}+\Delta h_c}$
A_I	i_o/i_i	$\dfrac{h_{21}}{1+h_{22}R_l}$	$\left[\dfrac{h_{fe}}{1+h_{oe}R_l}\right]\left[\dfrac{R_B}{R_B+R_i}\right]$	h_{fb}	$\dfrac{h_{fc}}{1+h_{oc}R_E} + \Delta h_c$
G	$|A_V A_I|$	$\dfrac{h_{21}^2 R_l}{(1+h_{22}R_l)(h_{11}+\Delta h R_l)}$	$|A_V A_I|$	$\dfrac{-h_{fb}^2}{h_{ib}G_l+\Delta h_b}$	$\dfrac{-h_{fc}^2}{h_{ic}G_l+\Delta h_c}$

$\Delta h_e = h_{ie}h_{oe} - h_{fe}h_{re}$

$\Delta h_b = h_{ib}h_{ob} - h_{fb}h_{rb}$

$\Delta h_c = h_{ic}h_{oc} - h_{fc}h_{rc}$

$\Delta h_b' = \dfrac{[h_{ib}(1+h_{ob}R_B) + (1-h_{rb})(1+h_{fb})R_B]h_{ob} - (h_{rb}+h_{ob}R_B)(h_{fb}-h_{ob}R_B)}{(1+h_{ob}R_B)^2}$

$Y_2' = \dfrac{1}{R_2} + \dfrac{1}{R_l} = G_l$

Common Emitter Equivalent Circuits

A. Transistor Circuits

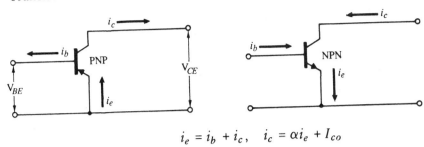

$$i_e = i_b + i_c, \quad i_c = \alpha i_e + I_{co}$$

B. T-Equivalent Circuits

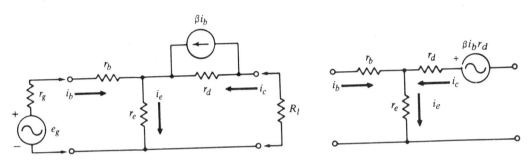

$$r_d = r_c(1 - \alpha) = \text{equivalent emitter–collector transresistance}$$

C. Hybrid-Equivalent Circuit

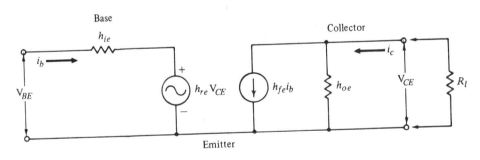

D. Simplified Gain Equations (see Tables 6-1 and 6-2)

$$A_V \approx -\frac{h_{fe}R_l}{h_{ie}} = \frac{h_{fb}R_l}{h_{ib}} = -\frac{\beta R_l}{r_b + r_e}$$

$$A_I = \beta = \frac{\alpha}{1 - \alpha} = \frac{\beta r_c(1 - \alpha)}{r_c(1 - \alpha) + R_l}$$

$$G_p = A_V A_I = \frac{\beta^2 R_l}{r_b + r_e}$$

Common Base Equivalent Circuits

A. Transistor Circuits

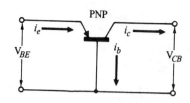

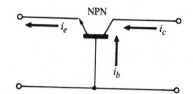

B. T-Equivalent Circuits

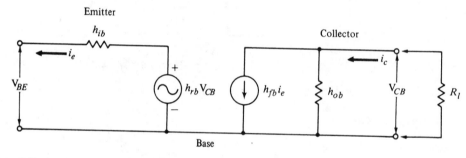

C. Hybrid-Equivalent Circuit

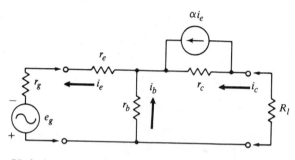

D. Simplified Gain Equations (see Tables 6-1 and 6-2)

$$A_V = \frac{\alpha R_l}{r_e + r_b(1 - \alpha)} \approx \frac{\alpha R_l}{r_e}$$

$$A_I = -\alpha$$

$$G_P = A_V A_I = \frac{\alpha^2 R_l}{r_e + r_b(1 - \alpha)} \approx \frac{\alpha^2 R_l}{R_e}$$

Common Collector (Emitter–Follower) Equivalent Circuits

A. Transistor Circuits

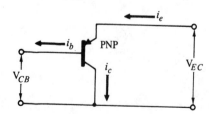

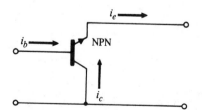

B. T-Equivalent Circuit

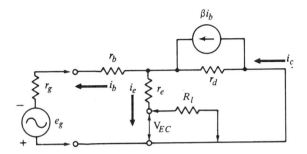

C. Hybrid-Equivalent Circuit

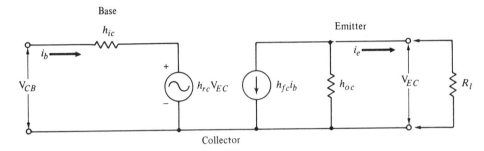

D. Simplified Gain Equations (see Tables 6-1 and 6-2)

$$A_V = \frac{-h_{fc}}{h_{ic}G_l + (h_{ic}h_{oc} - h_{rc}h_{fc})} = \frac{R_l}{R_l + r_e + r_b(1 - \alpha)} \approx 1$$

$$A_I = -(\beta + 1)$$

$$G_P = A_V A_I = \beta + 1 = \frac{1}{(1 - \alpha)}$$

PROBLEM 6-2. ONE-STAGE TRANSISTOR AMPLIFIER

Calculate the mid-frequency voltage gain A_V of the transistor circuit specified below.

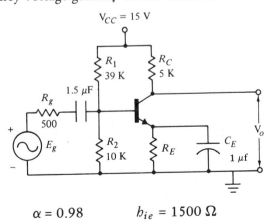

$$\alpha = 0.98 \qquad h_{ie} = 1500\ \Omega$$

$$r_e = 10\ \Omega \qquad h_{re} = 3 \times 10^{-4}$$

$$r_b = 1 \text{ K}\Omega \qquad h_{fe} = 49$$

$$r_c = 1.7 \text{ M}\Omega \qquad h_{oe} = 30 \times 10^{-6} \text{ } \mho$$

Solution:

Convert the above CE circuit to a hybrid-equivalent: *use thevenin equivalent circuit*

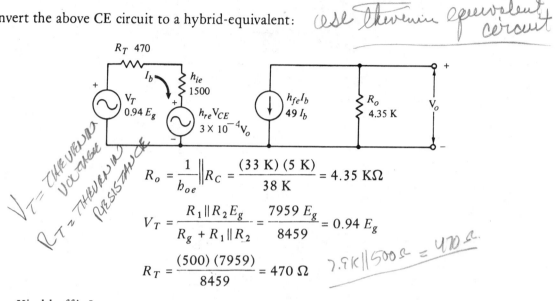

V_T = THEVENIN VOLTAGE

R_T = THEVENIN RESISTANCE

$$R_o = \frac{1}{h_{oe}} \Big\| R_C = \frac{(33 \text{ K})(5 \text{ K})}{38 \text{ K}} = 4.35 \text{ K}\Omega$$

$$V_T = \frac{R_1 \| R_2 E_g}{R_g + R_1 \| R_2} = \frac{7959 \, E_g}{8459} = 0.94 \, E_g$$

$$R_T = \frac{(500)(7959)}{8459} = 470 \, \Omega \qquad 7.9\text{K} \| 500\Omega = 470\Omega$$

From Kirchhoff's Laws:

$$V_o = -(4350)(49 \, I_b)$$

$$0.94 \, E_g = (470 + 1500) I_b + 3 \times 10^{-4} \, V_o$$

Solving for I_b :

$$0.94 \, E_g = 1970 I_b - 3 \times 10^{-4} \times 4350 \times 49 I_b$$

$$= I_b (1970 - 64) = 1906 I_b$$

$$I_b = \frac{0.94}{1906} E_g$$

(answer)
$$A_V = \frac{V_o}{E_g} = -\frac{(4350)(49)}{E_g} \left(\frac{0.94}{1906} E_g \right) = -105$$

Alternative Solution:

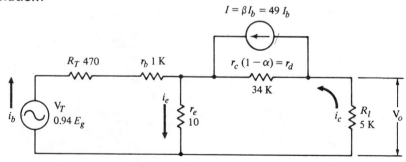

$$A_I = \frac{I_c}{I_b} = \frac{\beta r_c(1-\alpha)}{r_c(1-\alpha) + R_l} = \frac{49 \times 1.7 \times 10^6 \times 0.02}{1.7 \times 10^6 (0.02) + 5 \times 10^3} = \frac{1.67 \times 10^6}{3.9 \times 10^4} = 42.7$$

$$A_V = \frac{V_o}{E_g} = -0.94 \left[\frac{R_l}{R_i + R_T}\right] A_I$$

$$R_i = h_{ie} - \frac{h_{fe}h_{re}}{h_{oe} + G_l} = 1500 - \frac{(40)(3 \times 10^{-4})}{30 \times 10^{-6} + \frac{1}{5 \text{ K}}} = 1500 - 52 = 1448$$

$$A_V = -0.94 \left[\frac{5000}{1448 + 470}\right] 42.8 = -105 \qquad \text{(answer)}$$

Alternative Solution:

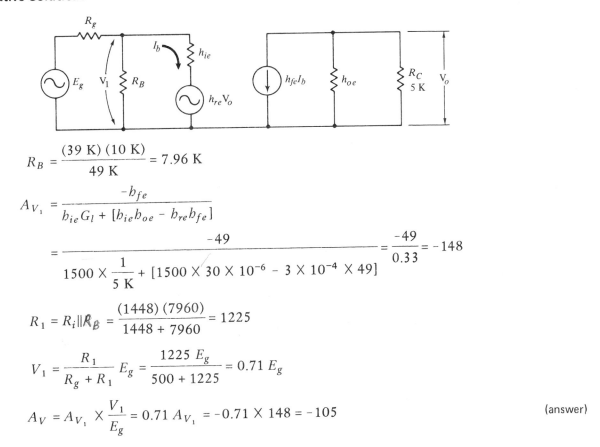

$$R_B = \frac{(39 \text{ K})(10 \text{ K})}{49 \text{ K}} = 7.96 \text{ K}$$

$$A_{V_1} = \frac{-h_{fe}}{h_{ie} G_l + [h_{ie}h_{oe} - h_{re}h_{fe}]}$$

$$= \frac{-49}{1500 \times \frac{1}{5 \text{ K}} + [1500 \times 30 \times 10^{-6} - 3 \times 10^{-4} \times 49]} = \frac{-49}{0.33} = -148$$

$$R_1 = R_i \| R_B = \frac{(1448)(7960)}{1448 + 7960} = 1225$$

$$V_1 = \frac{R_1}{R_g + R_1} E_g = \frac{1225 \; E_g}{500 + 1225} = 0.71 \; E_g$$

$$A_V = A_{V_1} \times \frac{V_1}{E_g} = 0.71 \; A_{V_1} = -0.71 \times 148 = -105 \qquad \text{(answer)}$$

If your answers are correct, go to Problem 6-3.
If your answers are not correct, review p. 153.

PROBLEM 6-3. COMMON BASE AMPLIFIER

For the common base amplifier circuit shown below, calculate the input impedance. Neglect all capacitances.

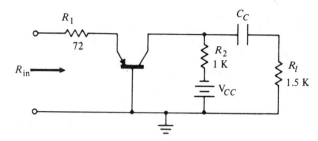

The transistor h-parameters are:

$$h_{ib} = 30\ \Omega$$

$$h_{rb} = 5 \times 10^{-4}$$

$$h_{fb} = -0.98$$

$$h_{ob} = 0.5 \times 10^{-6}\ \mho$$

Solution:

$$R_i = h_{ib} - \frac{h_{fb}h_{rb}}{Y'_2 + h_{ob}}, \qquad Y'_2 = \frac{1}{R_2} + \frac{1}{R_l} = \frac{1}{1000} + \frac{1}{1500} = \frac{1}{600}\ \mho$$

$$R_i = 30 - \frac{(-0.98)\,(5 \times 10^{-4})}{\dfrac{1}{600} + 0.5 \times 10^{-6}} = 30 + 0.29 = 30.29$$

(answer)
$$R_{in} = R_1 + R_i = 72 + 30.29 = 102.29\ \Omega$$

Alternative Solution:

$$R_i = \frac{\Delta h_b + h_{ib}\,Y'_2}{h_{ob} + Y'_2}, \qquad \Delta h_b = h_{ib}h_{ob} - h_{fb}h_{rb}$$

$$= (30)\,(0.5 \times 10^{-6}) - (-0.98)\,(5 \times 10^{-4}) = 5.05 \times 10^{-4}$$

(answer)
$$R_i = \frac{5.05 \times 10^{-4} + (30)\,(1.67 \times 10^{-3})}{0.5 \times 10^{-6} + 1.67 \times 10^{-3}} = 30.29, \quad R_{in} = 72 + 30.29 = 102.29\ \Omega$$

If your answers are correct, go to Problem 6-4.
If your answers are not correct, review p. 154.

PROBLEM 6-4. TWO-STAGE TRANSISTOR AMPLIFIER, COMMON EMITTER

The following information is given for the amplifier circuit shown below:

$$h_{ib} = 30\ \Omega, \quad h_{rb} = 5 \times 10^{-4}, \quad h_{fb} = -0.98, \quad h_{ob} = 0.5 \times 10^{-6}\ \mho$$

$$R_{b1} = 4.7\ \text{K}, \quad R_{l1} = 10\ \text{K}, \quad R_{b2} = 4.7\ \text{K}, \quad R_{l2} = 22\ \text{K}, \quad R_g = 500\ \Omega$$

Determine the voltage gain of each stage, overall power gain, interstage losses and pre–first stage losses, and total losses.

158

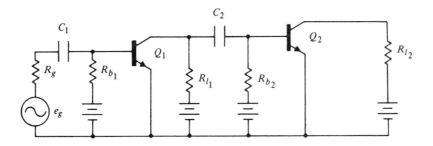

Solution:

The transistors are connected in common emitter configuration. Therefore, the h-parameters given in common base must be converted as follows:

$$h_{ie} = \frac{h_{ib}}{1 + h_{fb}} = \frac{30}{1 - 0.98} = 1500 \ \Omega$$

$$h_{re} = \frac{h_{ib}h_{ob}}{1 + h_{fb}} - h_{rb} = \frac{30 \times 0.5 \times 10^{-6}}{1 - 0.98} - 5 \times 10^{-4} = 2.5 \times 10^{-4}$$

$$h_{fe} = \frac{-h_{fb}}{1 + h_{fb}} = \frac{0.98}{1 - 0.98} = 49$$

$$h_{oe} = \frac{h_{ob}}{1 + h_{fb}} = \frac{0.5 \times 10^{-6}}{1 - 0.98} = 2.5 \times 10^{-5} \ \mho$$

This problem is best solved by breaking up the circuit and determining the gain of each stage as follows:

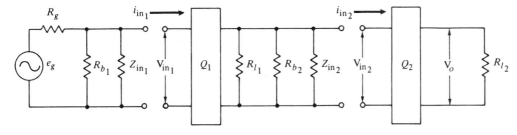

Capacitances are neglected.

The following two equations will be used in voltage gain and impedance calculations:

$$A_V = \frac{-h_{fe}}{h_{ie}G_l + (h_{ie}h_{oe} - h_{re}h_{fe})}$$

$$Z_{in} = h_{ie} - \frac{h_{re}h_{fe}}{h_{oe} + G_l}$$

Second stage voltage gain is:

$$A_{V2} = \frac{V_o}{V_{in\,2}} = \frac{-49}{(1500)(4.545 \times 10^{-5}) + (1500)(2.5 \times 10^{-5}) - (2.5 \times 10^{-4})(49)}$$

$$= -524.48 \qquad \text{(answer)}$$

First stage load impedance is the parallel combination of R_{L1}, R_{b2}, and Z_{in2}.

$$Z_{in2} = 1500 - \frac{(2.5 \times 10^{-4})(49)}{(2.5 \times 10^{-5}) + 4.545 \times 10^{-5}} = 1326 \ \Omega$$

Hence:

$$G'_{l1} = \frac{1}{R_{l1}} + \frac{1}{R_{b2}} + \frac{1}{Z_{in2}} = \frac{1}{10^4} + \frac{1}{4.7 \times 10^3} + \frac{1}{1.326 \times 10^3} = 1.067 \times 10^{-3} \ \mho$$

First stage voltage gain is:

$$A_{V1} = \frac{-49}{(1500)(1.067 \times 10^{-3}) + (1500)(2.5 \times 10^{-5}) - (2.5 \times 10^{-4})(49)}$$

(answer)
$$= -30.14$$

Total voltage gain $= A_V = (A_{V1})(A_{V2}) = (-30.14)(-524) = 15,808$

$$Z_{in1} = 1500 - \frac{(2.5 \times 10^{-4})(49)}{2.5 \times 10^{-5} + 1.067 \times 10^{-3}} = 1500 - 11.22 = 1489 \ \Omega$$

$$P_{in} = \frac{V_i^2}{R_{b1} \| Z_{in1}}, \quad R_{b1} \| Z_{in1} = \frac{(4700)(1489)}{4700 + 1489} = 1130.76 \ \Omega$$

$$P_{in} = \frac{V_i^2}{1130.76}$$

$$P_{out} = \frac{V_o^2}{R_{l2}} = \frac{V_o^2}{22,000}$$

Overall power gain is:

$$G = \frac{P_{out}}{P_{in}} = \left[\frac{V_o^2}{22,000}\right] \left[\frac{1130.76}{V_i^2}\right] = \frac{1130.76}{22,000} A_V^2 = \frac{1130.76}{22,000} (15,808)^2$$

(answer)
$$= 1.284 \times 10^7$$

In db,

(answer)
$$G_{db} = 10 \log G = 10(7 + 0.109) = 71.09 \ db$$

Interstage power loss is that power lost in the parallel combination of R_{l1} and R_{b2}.

$$R_{l1} \| R_{b2} = \frac{(10 \ K)(4.7 \ K)}{10 \ K + 4.7 \ K} = 3200 \ \Omega$$

(answer)
$$P_{IL} = \frac{V_{in2}^2}{3200} = \frac{(30.14 V_{in1})^2}{3200} = 0.28 V_{in1}^2$$

Pre-first stage loss is that power lost in R_{b1}:

(answer)
$$P_{PFSL} = \frac{V_{in1}^2}{R_{b1}} = \frac{V_{in1}^2}{4700} = 2.13 \times 10^{-4} V_{in1}^2$$

160

Total losses in terms of e_g are calculated as follows: Make a Thevenin equivalent circuit out of the pre-first stage circuit:

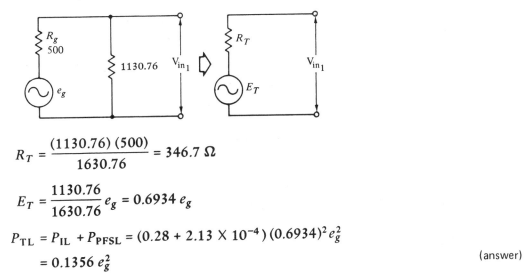

$$R_T = \frac{(1130.76)(500)}{1630.76} = 346.7 \ \Omega$$

$$E_T = \frac{1130.76}{1630.76} e_g = 0.6934 \ e_g$$

$$P_{TL} = P_{IL} + P_{PFSL} = (0.28 + 2.13 \times 10^{-4})(0.6934)^2 e_g^2$$

$$= 0.1356 \ e_g^2$$

(answer)

If your answers are correct, go to Problem 6-5.
If your answers are not correct, review pp. 151 and 153.

PROBLEM 6-5.   TWO-STAGE VOLTAGE GAIN, h-PARAMETERS

The h=parameters and bias resistor values are given for the two-stage transistor amplifier that follows. Assume all capacitors are short circuits at signal frequency. Determine the voltage gain, V_2/V_s.

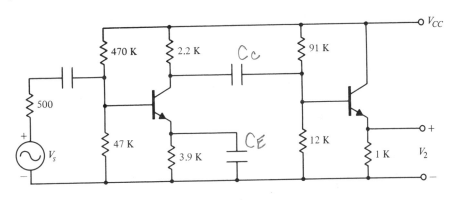

$h_{ie} = 1500 \ \Omega$

$h_{re} = 3 \times 10^{-4}$

$h_{fe} = 49$

$h_{oe} = 30 \times 10^{-6} \ \mho$

$h_{ic} = 1500 \ \Omega$

$h_{rc} = 1$

$h_{fc} = -50$

$h_{oc} = 30 \times 10^{-6} \ \mho$

Solution:

Notice that the first stage is CE and the second stage is CC (emitter follower). Draw an equivalent circuit of the first stage.

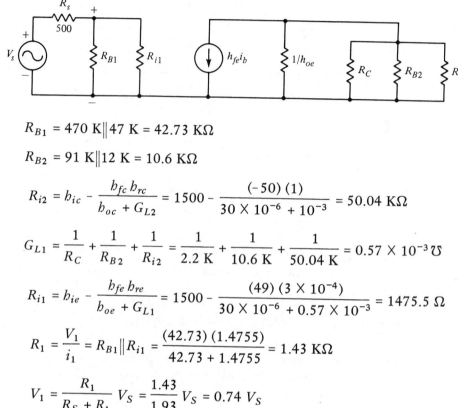

$$R_{B1} = 470 \text{ K} \| 47 \text{ K} = 42.73 \text{ K}\Omega$$

$$R_{B2} = 91 \text{ K} \| 12 \text{ K} = 10.6 \text{ K}\Omega$$

$$R_{i2} = h_{ic} - \frac{h_{fc} h_{rc}}{h_{oc} + G_{L2}} = 1500 - \frac{(-50)(1)}{30 \times 10^{-6} + 10^{-3}} = 50.04 \text{ K}\Omega$$

$$G_{L1} = \frac{1}{R_C} + \frac{1}{R_{B2}} + \frac{1}{R_{i2}} = \frac{1}{2.2 \text{ K}} + \frac{1}{10.6 \text{ K}} + \frac{1}{50.04 \text{ K}} = 0.57 \times 10^{-3} \, \mho$$

$$R_{i1} = h_{ie} - \frac{h_{fe} h_{re}}{h_{oe} + G_{L1}} = 1500 - \frac{(49)(3 \times 10^{-4})}{30 \times 10^{-6} + 0.57 \times 10^{-3}} = 1475.5 \, \Omega$$

$$R_1 = \frac{V_1}{i_1} = R_{B1} \| R_{i1} = \frac{(42.73)(1.4755)}{42.73 + 1.4755} = 1.43 \text{ K}\Omega$$

$$V_1 = \frac{R_1}{R_S + R_1} V_S = \frac{1.43}{1.93} V_S = 0.74 \, V_S$$

$$R_{L1} = \frac{1}{G_{L1}} = \frac{1}{0.57 \times 10^{-3}} = 1754.4 \, \Omega$$

The voltage gain of the first stage (CE) is:

$$A_{V1} = \frac{-h_{fe}}{h_{ie} G_{L1} + \Delta_{he}}$$

$$= \frac{-49}{(1500)(0.57 \times 10^{-3}) + [(1500)(30 \times 10^{-6}) - (49)(3 \times 10^{-4})]} = 55.35$$

The voltage gain of the second stage (CC) is:

$$A_{V2} = \frac{-h_{fc}}{h_{ic} G_{L2} + \Delta_{hc}}$$

$$= \frac{-(-50)}{(1500)(10^{-3}) + [(1500)(30 \times 10^{-6}) - (50)(1)]} = \frac{50}{51.55} = 0.97$$

$$V_2 = V_1 A_{V1} A_{V2} = (0.74\ V_s)\ (-55.35)\ (0.97) = -39.73\ V_s$$

$$\frac{V_2}{V_s} = -39.73 \qquad\qquad \text{(answer)}$$

If your solution is correct, go on to Problem 6-6.
If your solution is not correct, review pp. 151–155.

PROBLEM 6-6.   CURRENT GAIN, Z_{OUT}, h-PARAMETERS

The h-parameters for the single-stage transistor shown in the following circuit are:

$b_{ie} = 1400\ \Omega$ $\qquad\qquad$ $b_{re} = 4 \times 10^{-4}$
$b_{fe} = 50$ $\qquad\qquad\qquad$ $b_{oe} = 25 \times 10^{-6}\ \mho$

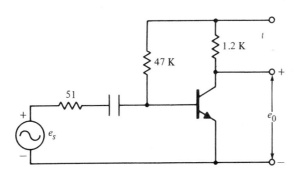

Calculate the current gain and output impedance.

Solution:

$$R_i = b_{ie} - \frac{b_{re}\,b_{fe}\,R_L}{1 + b_{oe}\,R_L} = 1400 - \frac{(4 \times 10^{-4})\,(50)\,(1.2 \times 10^3)}{1 + (25 \times 10^{-6})\,(1.2 \times 10^3)} = 1376.7\ \Omega$$

$$A_I = \frac{b_{fe}}{1 + b_{oe}\,R_L}\left[\frac{R_B}{R_B + R_i}\right] = \frac{50}{1 + (25 \times 10^{-6})\,(1.2 \times 10^3)}\left[\frac{47}{47 + 1.3767}\right]$$

$$= 47.16 \qquad\qquad \text{(answer)}$$

$$Y_{\text{out}} = b_{oe} - \frac{b_{fe}\,b_{re}}{R_s + b_{ie}} = 25 \times 10^{-6} - \frac{(50)\,(4 \times 10^{-4})}{51 + 1400} = 1.12 \times 10^{-5}\ \mho$$

$$Z_{\text{out}} = \frac{1}{Y_{\text{out}}} = 89.231\ \text{K}\Omega \qquad\qquad \text{(answer)}$$

If your answers are correct, go on to Problem 6-7.
If your answers are not correct, review pp. 151–155.

PROBLEM 6-7. TRANSISTOR AMP—UPPER AND LOWER CUTOFF FREQUENCY

The following circuit represents one stage of an RC-coupled amplifier:

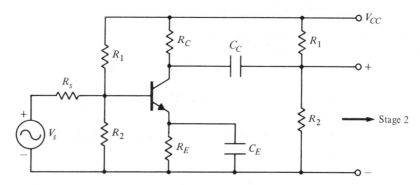

$h_{ie} = 1800 \ \Omega$
$h_{re} = 3 \times 10^{-4}$
$h_{fe} = 50$
$h_{oe} = 5 \times 10^{-6} \ \mho$

beta cutoff frequency, $f_B = 2.5$ MHz
$R_B = R_1 \| R_2 = 4.7$ K
$R_E = 50 \ \Omega$
$C_p = 620$ pf
$R_s = 2200 \ \Omega$

a. Calculate R_C to yield an upper cutoff frequency, f_2, of 300 KHz.
b. Determine the mid-frequency current gain.
c. Determine the values of C_C and C_E to yield a lower cutoff frequency, f_1, of 200 Hz.

Solution:

a. Since f_β is well above f_2, the output resistance becomes:

$$R_o = \frac{1}{2\pi f_2 C_p} = \frac{1}{2\pi \times 300 \times 10^3 \times 620 \times 10^{-12}} = 856 \ \Omega$$

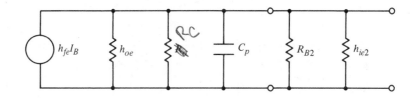

$$G_o = \frac{1}{R_o} = \frac{1}{856} = \frac{1}{R_C} + h_{oe} + \frac{1}{R_{B2}} + \frac{1}{h_{ie2}}$$

$$= \frac{1}{R_C} + 5 \times 10^{-6} + \frac{10^{-3}}{4.7} + \frac{1}{1800} = \frac{1}{R_C} + 7.73 \times 10^{-4}$$

$$\frac{1}{R_C} = \frac{1}{856} - 7.73 \times 10^{-4} = 3.95 \times 10^{-4}$$

(answer) $R_C = 2530 \ \Omega$

b.

$$A_{I\,\text{mid}} = -\frac{h_{fe}R_o}{h_{ie}} = -\frac{50 \times 856}{1800} = -23.78 \qquad \text{(answer)}$$

c.

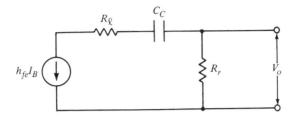

$$R_\ell = \frac{R_C/h_{oe}}{R_C + 1/h_{oe}} = \frac{2530/5 \times 10^{-6}}{2530 + \dfrac{1}{5 \times 10^{-5}}} = 2498.4\ \Omega$$

$$R_r = \frac{R_B h_{ie}}{R_B + h_{ie}} = \frac{4.7 \times 10^3 \times 1800}{4.7 \times 10^3 + 1800} = 1301.5\ \Omega$$

$$C_C = \frac{1}{2\pi f_1(R_\ell + R_r)} = \frac{1}{2\pi \times 200\,(2498.4 + 1301.5)} = 0.21\ \mu f \qquad \text{(answer)}$$

$$C_E = \frac{1 + h_{fe}}{2\pi f_1 R_s} = \frac{1 + 50}{2\pi \times 200 \times 2200} = 18.4\ \mu f \qquad \text{(answer)}$$

If your answers are correct, go on to Problem 6-8.

If your answers are not correct, review pp. 151–155.

PROBLEM 6-8. DARLINGTON/TRANSIENT PROBLEM

Determine the voltage across the capacitor in the darlington amplifier circuit below 0.25 second after the switch is open.

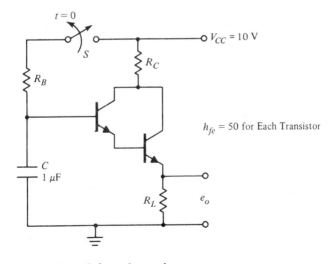

Assume the switch has been closed for a long time.

Solution:

Assume capacitor is initially charged to $V_{CC} = 10$ volts. Current gain of a darlington is $\beta^2 = 50^2 = 2500$. An equivalent circuit with R_L reflected into the input circuit is:

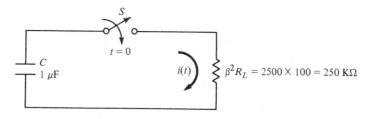

$$\tau = RC = 250 \times 10^3 \times 10^{-6} = 0.25 \text{ sec.}$$

$$V_C(t) = 10^{e^{-t/\tau}} = 10\, e^{-4t}$$

For $t = 0.25$ seconds, $V_C = 10e^{-1} = 3.7$ volts   (answer)

If your answer is correct, go on to the next section.
If your answer is not correct, review pp. 12–16.

BIASING AND STABILITY

All transistors in linear circuit applications require some form of biasing in order to establish collector–base–emitter voltage and current relationships at the *operating point* (quiescent point, Q-point, no-signal point) of the circuit. The actual circuit configuration and bias circuit values are selected on the basis of dynamic current conditions (desired output voltage swing, expected input signal level, class of operation, desired gain, supply voltages, transistor type, input and output impedances, etc.). The basic bias network must maintain the desired base current in the presence of temperature and frequency changes (referred to as bias stability).

The following problems illustrate biasing techniques and stability factor calculations.

PROBLEM 6-9.   TRANSISTOR CURVES AND LOAD LINE

A common emitter transistor circuit is shown below.

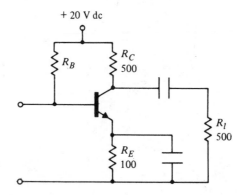

The transistor characteristic curves are shown below. It is desired to operate the transistor at V_{CE} = 10 V dc when no signal is applied.

- a. Draw the dc load line.
- b. Calculate R_B.
- c. Draw the ac load line.
- d. Calculate h_{fe} and h_{oe}.

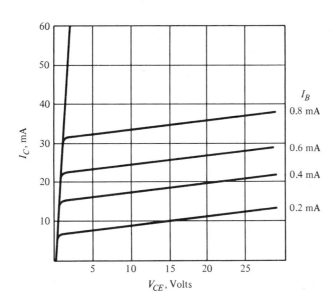

Solution:

a. $I_C \Big|_{V_{CE}=0} = \dfrac{20}{(100 + 500)} = 33.3$ mA

$V_{CE} \Big|_{I_C=0} = 20$ V

Thus, the dc load line intersects the ordinate at 33.3 mA and the abscissa at 20 V.   (answer)

b. The Q-point is indicated on the dc load line at V_{CE} = 10 V and I_C = 16.7 mA. At this point, I_B = 0.4 mA. The three transistor voltages may be determined as follows:

$$V_C = 20 - (500)(16.7 \times 10^{-3}) = 20 - 8.3 = 11.7 \text{ V}$$

$$V_E = (100)(16.7 \times 10^{-3}) = 1.7 \text{ V}$$

V_B is one diode drop (assume 1 V) above V_E.

$$\therefore V_B = 1.7 + 1 = \cancel{2.7 \text{ V}} \quad 2.3V$$
$$0.6$$

Thus:

$$R_B = \frac{(20 - \cancel{2.7})^{2.3}}{0.4 \times 10^{-3}} = \cancel{43.25} \text{ K}\Omega \quad 44.25 K\Omega.$$ (answer)

c. R_{ac} = 500‖500 = 250 Ω (answer)

The ac load line has the slope, $-1/R_a = -1/250 = -4$ mA/V and passes through Q.

167

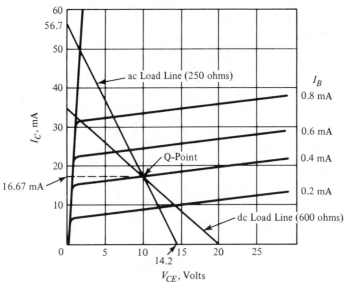

d. At $V_{CE} = 10$ volts,

(answer)

$$b_{fe} = \frac{\Delta I_C}{\Delta I_B}\bigg|_{\Delta V_{CE}=0} = \frac{24.4 - 9}{(0.6 - 0.2)} = \frac{15.4}{0.4} = 38.5$$

(answer)

$$b_{oe} = \frac{\Delta I_C}{\Delta V_{CE}}\bigg|_{\Delta I_B=0} = \frac{(17.7 - 15.7)}{15 - 5} \times 10^{-3} = \frac{2 \times 10^{-3}}{10} = 0.2 \times 10^{-3} \ \mho$$

If your answers are correct, go on to Problem 6-10.

If your answers are not correct, review p. 115.

PROBLEM 6-10. TRANSISTOR STABILITY

The silicon NPN common emitter circuit configuration shown below has the following given information:

Q-point: $V_{CE} = 5$ volts, $I_c = 1$ mA

$$R_1 \| R_2 = R_B = 4000 \ \Omega$$

C_E and C_C are negligible at the operating frequency but block dc currents.

$$V_{BE} = 0.7 \text{ volt}$$

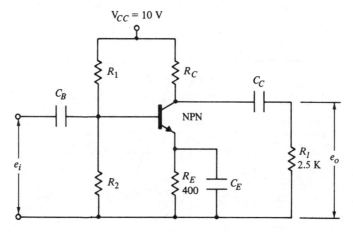

168

a. Calculate R_C, R_1, and R_2.

b. Calculate the stability factor S.

c. Determine the new Q-point if I_{CO} increases by 50 μamps.

Solution:

a. $I_C(R_E + R_C) = V_{CC} - V_{CE}$

$$I_E = \frac{I_C}{\alpha} = \frac{10^{-3}}{0.98} = 1.02 \times 10^{-3}$$

$$V_E = 400\, I_E = 0.408 \text{ volt}$$

$$R_C = \frac{V_{CC} - V_C}{I_C} = \frac{10 - (5 + 0.408)}{10^{-3}} = 4592\ \Omega \qquad \text{(answer)}$$

$$V_B = V_E + V_{BE} = 0.408 + 0.7 = 1.108 \text{ volts}$$

$$\frac{10}{R_1 + R_2} = \frac{1.108}{R_2}$$

$$10R_2 = 1.108\,(R_1 + R_2)$$

$$9.03R_2 = R_1 + R_2$$

$$R_1 = 8.03\,R_2$$

$$\frac{R_1 R_2}{R_1 + R_2} = 4000 = \frac{8.03\,R_2^2}{9.03\,R_2} = 0.89\,R_2$$

$$R_2 = \frac{4000}{0.89} = 4498\ \Omega \qquad \text{(answer)}$$

$$R_1 = 8.03\,R_2 = 36{,}120\ \Omega \qquad \text{(answer)}$$

b. Transistor stability is a function of collector leakage current (I_{CO}) change with temperature, and its effect on I_C. Stability factor, S, may be defined by the following equations:

$$S = \frac{\Delta I_C}{\Delta I_{CO}} = \frac{1 + R_E/R_B}{1 - \alpha + R_E/R_B} = \frac{R_B + R_E}{R_B(1 - \alpha) + R_E} = \frac{\beta + 1}{1 + \dfrac{\beta R_E}{R_E + R_B}}$$

It is desirable to have as low a value of S as possible. For this problem:

$$S = \frac{R_B + R_E}{R_B(1 - \alpha) + R_E} = \frac{4000 + 400}{4000(1 - 0.98) + 400} = 9.17 \qquad \text{(answer)}$$

c. $\Delta I_C = S(\Delta I_{CO}) = (9.17)(50 \times 10^{-6}) = 4.58 \times 10^{-4}$ amp

new $I_C = 10^{-3} + 0.458 \times 10^{-3} = 1.458$ mA

$I_C(R_E + R_C) = V_{CC} - V_{CE}$

$V_{CE} = V_{CC} - I_C(R_E + R_C) = 10 - 1.458 \times 10^{-3}(400 + 4592)$

$\qquad = 2.72$ volts $\qquad \text{(answer)}$

Thus, it is seen that the Q-point has shifted considerably.

A PROGRAMMED REVIEW FOR ELECTRICAL ENGINEERING

If your answers are correct, go on to Problem 6-11.
If your answers are not correct, review the preceding solution.

PROBLEM 6-11. TRANSISTOR SPECS, STABILITY

The single-stage transistor amplifier that follows has the following h-parameters:

$h_{ie} = 2500\ \Omega$
$h_{fe} = 120$
$h_{oe} = 10^{-5}\ \mho$
$h_{FE} = 145$

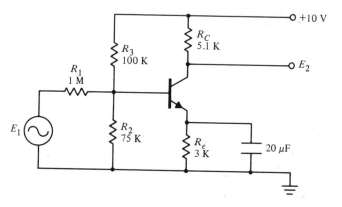

Calculate the voltage gain E_2/E_1, current gain, and stability factor.

Solution:

Drawing an equivalent circuit:

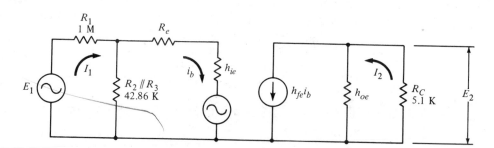

Assume $h_{re} = 0$ since it is not given.

$$i_b = I_1 \frac{42.86}{42.86 + h_{ie} + R_e} = 0.886\ I_1$$

$$I_2 = 120\ i_b \left(\frac{100}{5.1 + 100}\right) = 120\ (0.945\ I_1)\ (0.9515) = -101.1\ I_1$$

0.886   = 107.9

$$A_I = \frac{I_2}{I_1} = -101.1 \qquad \text{(answer)}$$

170

$$E_2 = I_2 \, (5.1 \text{ K}) = (101.1 \, I_1) \, (5.1 \text{ K}) = 515,600$$

$$E_1 = 10^6 \, I_1$$

$$A_V = \frac{E_2}{E_1} = \frac{515,600}{10^6} = 0.5156 \qquad \text{(answer)}$$

$$S = \frac{\beta + 1}{1 + \dfrac{\beta R_E}{R_E + R_B}} = \frac{145 + 1}{1 + \dfrac{(145)\,(3)}{3 + 42.86}} = 13.92 \qquad \text{(answer)}$$

If your answers are correct, go on to the next section.
If your answers are not correct, review pp. 168 and 169.

FIELD EFFECT TRANSISTORS

The field effect transistor (FET) is different from the bipolar junction transistor in the following important ways:

1. FET operation depends upon the flow of majority carriers only. Therefore, it is a unipolar device.
2. It has a much higher input resistance, typically many megohms.
3. It is less noisy than a bipolar transistor.
4. It is simpler to fabricate and occupies less area when used in LSI technology.
5. Thermal runaway is not a problem.

The main disadvantage is the relatively small gain-bandwidth product.
The typical circuit symbol for an *n*-channel FET is as follows:

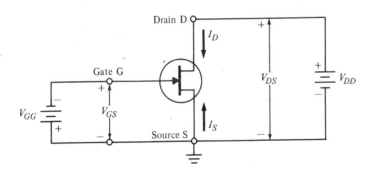

Low-frequency and high-frequency (including node capacitors), small-signal FET models are as follows:

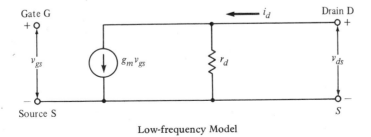

Low-frequency Model

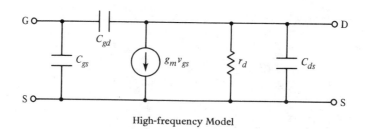

High-frequency Model

Because of the internal capacitances of a FET, feedback exists between output and input circuits, and voltage amplification drops off rapidly with frequency. Typical parameters for junction and MOS FETs are given in the table below:

Parameter	JFET	MOSFET
g_m	0.1 – 10 ma/volt	0.1 – 50 ma/volt
r_d	0.1 – 1 mΩ	1 – 50 KΩ
C_{ds}	0.1 – 1 pf	0.1 – 1 pf
C_{gs}, C_{gd}	1 – 10 pf	1 – 10 pf
r_{gs}	$>10^8$ Ω	$>10^{10}$ Ω
r_{gd}	$>10^8$ Ω	$>10^{14}$ Ω

The FET may be connected in common source, CS, or common drain, CD, configuration, as shown below. No biasing is shown.

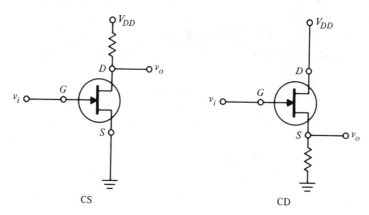

CS   CD

Compared to a bipolar transistor, CS is equivalent to CE, and CD is equivalent to CC. The circuit below shows a single power supply biased FET circuit in the CS configuration.

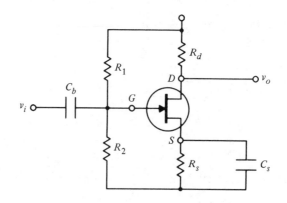

PROBLEM 6-12.   FIELD EFFECT TRANSISTOR AMPLIFIER

The circuit of a two-stage FET amplifier is shown below.   Each FET has the following characteristics:

$$g_m = 3 \times 10^{-3} \ \mho, \quad r_d = 6.8 \ K\Omega, \quad R_d = 10 \ K\Omega, \quad R_g = 47 \ K\Omega$$

$$C_g = 0.01 \ \mu f, \qquad C_{sh} = 40 \ pf$$

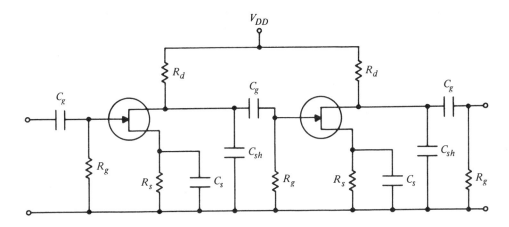

Determine the overall midband voltage gain, lower 3db frequency, and upper 3db frequency.

Solution:

A junction FET has properties analogous to a vacuum tube.   Its voltage gain is given by the formula:

$$A_V = -g_m R$$

In this case, $R = r_d \| R_d \| R_g$

$$\frac{1}{R} = \frac{1}{6.8 \times 10^3} + \frac{1}{10^4} + \frac{1}{47 \times 10^3} = 2.683 \times 10^{-4}$$

$$R = 3727 \ \Omega$$

$$A_V = -3 \times 10^{-3} \times 3727 = 11.18 \text{ per stage}$$

Overall gain is: $A_{VT} = A_V^2 = 11.18^2 = 125$ (answer)

In db, $A_{VT} = 20 \log 125 = 42 \text{ db}$ (answer)

Lower 3 db frequency, f_1, for each stage is:

$$f_1 = \frac{1}{2\pi R_1 C_g},$$

where

$$R_1 = R_g + r_d \parallel R_d$$

$$= 47 \times 10^3 + \frac{(6.8 \times 10^3)(10^4)}{16.8 \times 10^3}$$

$$= 51.05 \text{ K}\Omega$$

$$f_1 = \frac{1}{2\pi \times 51.05 \times 10^3 \times 10^{-8}} = 311.76 \text{ Hz}$$

Overall 3 db frequency is:

(answer)

$$f_{1n} = \frac{f_1}{\sqrt{2^{1/n} - 1}} = \frac{311.76}{\sqrt{2^{1/2} - 1}} = 484.41 \text{ Hz}$$

Upper 3 db frequency, f_2, for each stage is:

$$f_2 = \frac{1}{2\pi R C_{sb}} = \frac{1}{2\pi \times 3727 \times 40 \times 10^{-12}} = 1.07 \text{ MHz}$$

Overall 3 db frequency is:

(answer)

$$f_{2n} = f_2 \sqrt{2^{1/2} - 1} = 1.07 \times 10^6 \sqrt{2^{1/2} - 1} = 687 \text{ KHz}$$

If your answers are correct, go to the next section.
If your answers are not correct, review the above solution.

VACUUM TUBES

The following is a brief summary of vacuum tube characteristics and equivalent circuits.

Vacuum Tube Characteristics

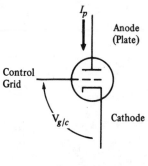

Triode Schematic

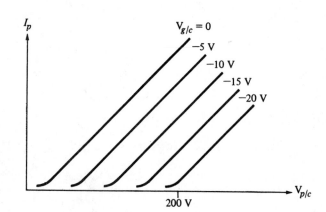

Black Box

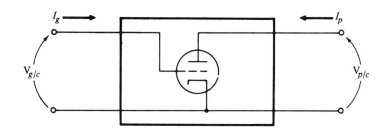

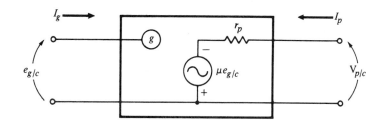

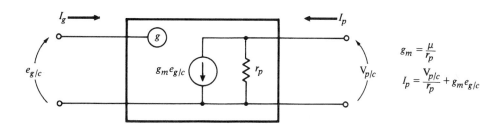

$$g_m = \frac{\mu}{r_p}$$

$$I_p = \frac{V_{p/c}}{r_p} + g_m e_{g/c}$$

Circuit Model

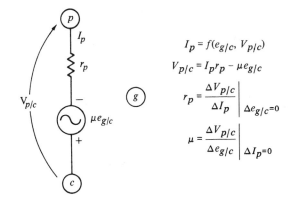

$$I_p = f(e_{g/c}, V_{p/c})$$

$$V_{p/c} = I_p r_p - \mu e_{g/c}$$

$$r_p = \frac{\Delta V_{p/c}}{\Delta I_p}\bigg|_{\Delta e_{g/c}=0}$$

$$\mu = \frac{\Delta V_{p/c}}{\Delta e_{g/c}}\bigg|_{\Delta I_p=0}$$

175

Characteristics with Load Line

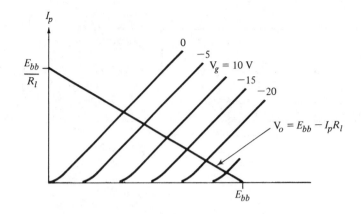

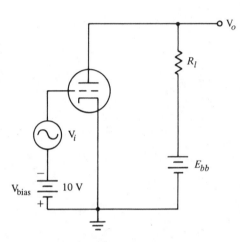

$$V_o = -I_p R_l$$

$$I_p = \frac{\mu e_g}{r_p + R_l}$$

$$V_o = -\mu V_i \frac{R_l}{R_l + r_p}$$

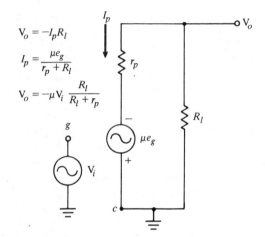

Triode Applications

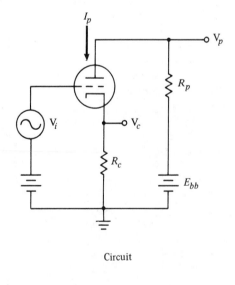

Circuit

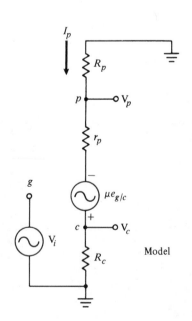

Model

$$V_p = -I_p R_p$$

$$V_C = +I_p R_C$$

$$\mu e_{c/g} = I_p(R_C + R_p + r_p) = (V_i - I_p R_C)\mu$$

$$\therefore \mu V_i - \mu I_p R_C = I_p(R_C + R_p + r_p)$$

$$I_p = \frac{\mu V_i}{R_p + r_p + (1 + \mu)R_C}$$

$$V_p = \frac{-\mu V_i r_p}{R_p + r_p + (1 + \mu) R_C}$$

$$V_C = \frac{\mu V_i R_C}{R_p + r_p + (1 + \mu)R_C}$$

Output Impedance

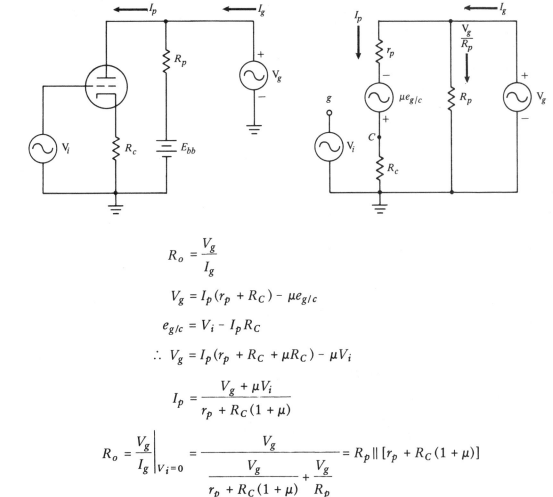

$$R_o = \frac{V_g}{I_g}$$

$$V_g = I_p(r_p + R_C) - \mu e_{g/c}$$

$$e_{g/c} = V_i - I_p R_C$$

$$\therefore V_g = I_p(r_p + R_C + \mu R_C) - \mu V_i$$

$$I_p = \frac{V_g + \mu V_i}{r_p + R_C(1 + \mu)}$$

$$R_o = \frac{V_g}{I_g}\bigg|_{V_i=0} = \frac{V_g}{\dfrac{V_g}{r_p + R_C(1 + \mu)} + \dfrac{V_g}{R_p}} = R_p \| [r_p + R_C(1 + \mu)]$$

177

PROBLEM 6-13. VACUUM TUBE AMPLIFIER

The pentode amplifier circuit below has the following mid-band characteristics:

$$g_m = 5 \times 10^{-3} \; \mho$$

$$\mu = 5000$$

$$R_l = 30 \; K\Omega$$

$$r_p = 1 \; M\Omega$$

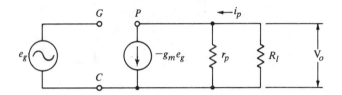

Determine the voltage gain.

Solution:

$$A_V = \frac{v_o}{e_g}, \quad v_o = -g_m e_g \left[\frac{R_l r_p}{R_l + r_p}\right]$$

$$A_V = -g_m \left[-\frac{R_l r_p}{R_l + r_p}\right] = -5 \times 10^{-3} \left[\frac{(30 \times 10^3)(10^6)}{30 \times 10^3 + 10^6}\right] = -145.6 \quad \text{(answer)}$$

If your answer is correct, go to the next section.
If your answer is not correct, review the above solution.

OPERATIONAL AMPLIFIERS

General

A typical operational amplifier circuit application is shown below:

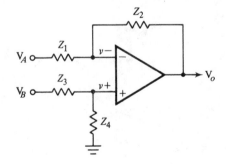

Parameter	Ideal	μA741
A_v	∞	10^5 V/V
R_i	∞	$10^5 \; \Omega$
i_-, i_+	0	10^{-7} A

Since the amplifier is ideal, $i- = 0 = i+$ and $v- = v+$ is forced by the amplifier.

$$V_+ = \frac{Z_4}{Z_3 + Z_4} V_B \quad \text{(voltage divider method)}$$

$$V_- = V_A \frac{Z_2}{Z_1 + Z_2} + V_o \frac{Z_1}{Z_1 + Z_2} \quad \text{(superposition method)}$$

$$\frac{Z_4}{Z_3 + Z_4} V_B = \frac{Z_2 V_A + Z_1 V_o}{Z_1 + Z_2}, \quad V_o = \frac{Z_1 + Z_2}{Z_1} \left[V_B \frac{Z_4}{Z_3 + Z_4} - V_A \frac{Z_2}{Z_1 + Z_2} \right]$$

$$\therefore V_o = V_B \left[\frac{Z_1 + Z_2}{Z_1} \right] \left[\frac{Z_4}{Z_3 + Z_4} \right] - V_A \frac{Z_2}{Z_1}$$

Special Cases

Shown below is a collection of operational amplifier circuit applications:

1. Differential Amplifier.

$$\frac{Z_2}{Z_1} = \frac{Z_4}{Z_3}, \quad V_o = (V_B - V_A) \frac{Z_2}{Z_1}$$

2. Inverting Amplifier.

$$V_0 = -V_A \left[\frac{Z_2}{Z_1} \right]$$

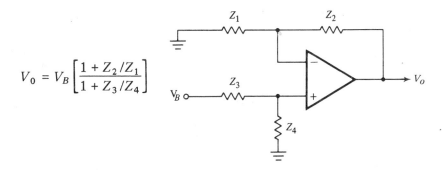

3. Noninverting Amplifier.

$$V_0 = V_B \left[\frac{1 + Z_2/Z_1}{1 + Z_3/Z_4} \right]$$

4. Voltage Follower.

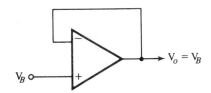

5. Low Pass Inverting.

$$\frac{V_o}{V_i}(s) = \frac{-R_f/R_i}{1 + sR_fC_f}$$

1st order lag at $\quad \omega = \dfrac{1}{R_fC_f}$

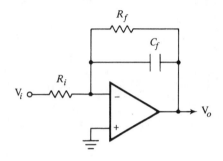

6. Integrator.

$$\frac{V_o}{V_i}(s) = \frac{-1}{sRC}; \quad V_o(t) = -\frac{1}{RC}\int V_i\,dt$$

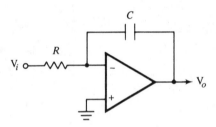

7. Differentiator.

$$\frac{V_o}{V_i}(s) = -sRC; \quad V_o(t) = -RC\,\frac{dV_i(t)}{dt}$$

(noise sensitive due to input C)

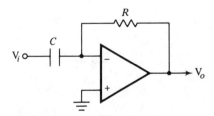

For additional op amp circuits, see references 40 and 41 in the bibliography.

PROBLEM 6-14. OPERATIONAL AMPLIFIER

Determine the output voltage for the circuit shown below if the input is 0.2 sin ωt volts.

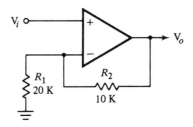

Solution:

This circuit is a noninverting amplifier. Output voltage is given by the formula:

$$V_o = V_i \left[1 + \frac{R_2}{R_1} \right]$$

$$\therefore \ V_o = 0.2 \sin \omega t \ (1 + 0.5) = 0.3 \sin \omega t \text{ volts}$$ (answer)

If your answer is correct, go on to Problem 6-15.
If your answer is not correct, review p. 179.

PROBLEM 6-15 AC/DC CONVERTER

For the ac/dc converter circuit shown below, determine the values for R_1, R_2, and R_3 that will provide a dc voltage equal to the RMS value of an input sine wave ac voltage. Assume $V_i = V_m \sin \omega t$, where $V_m \leqslant 18$ v, and the ideal op amps limit at ± 15 volts.

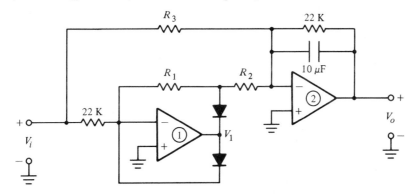

Solution:

Op amp 1 is an inverting rectifier. Op amp 2 is an inverting summer. For a sinusoid, RMS value = 0.707 V_m. Select R_1 such that $V_1 = 15$ V when $V_i = 18$ V.

$$V_1 = -V_i \frac{R_1}{22 \text{ K}} = -18 \frac{R_1}{22 \text{ K}} = -15$$

$$R_1 = \frac{15}{18} \times 22 \text{ K} = 18.3 \text{ K}$$ (answer)

Rectified sine wave:

$$V_m K_1\left[\frac{1}{\pi} + \frac{1}{2}\sin \omega t - \frac{2}{\pi}\left(\frac{\cos 2\omega t}{3}\right) + \cdots\right]$$

$$K_1 = \frac{R_1}{22\text{ K}} = 0.833$$

Summer:

$$V_o = -\left[\frac{22\text{ K}}{R_3}(V_m \sin \omega t) + V_1\right]$$

$$V_o = -\frac{22\text{ K}}{R_3}V_m \sin \omega t + \frac{22\text{ K}}{R_2}(0.833V_m)\left[\frac{1}{\pi} + \frac{1}{2}\sin \omega t - \frac{2}{\pi}\cdots\right]$$

$$= \frac{22\text{ K}}{R_2}(0.833V_m)\frac{1}{\pi} + \left[\frac{22\text{ K}}{2R_2}(0.833V_m) - \frac{22\text{ K}}{R_3}V_m\right]\sin \omega t + \cdots$$

Choose R_2 and R_3 such that the sin term cancels out,

$$\text{i.e., } \frac{22\text{ K}}{2R_2}(0.833V_m) = \frac{22\text{ K}}{R_3}V_m$$

$$R_3 = 2.4R_2$$

$$V_o = \frac{22\text{ K}}{R_2}(0.833V_m)\frac{1}{\pi} + \text{harmonic terms}$$

Assuming harmonics are negligible because of the filter:

$$V_o = 0.707V_m = \frac{22\text{ K}}{R_2}(0.833V_m)\frac{1}{\pi}$$

(answer)
$$R_2 = \frac{22\text{ K}(0.833)}{0.707\,\pi} = 8.25\text{ K}$$

(answer)
$$R_3 = 2.4R_2 = 19.8\text{ K}$$

If your answers are correct, go on to the next section.
If your answers are not correct, review Dobkin[40] and check your calculations.

AMPLIFIER CLASS

Amplifier stages can be classified according to *class* of operation in response to a sinusoidal input, as indicated on p. 183.

182

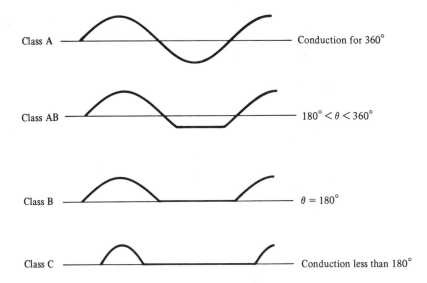

In general, class A is the most linear (contains the least distortion) and is the least efficient. Class C is the other extreme.

If the input to an amplifier is a pure sine wave of single frequency, but the output contains higher harmonics (which can be expressed in Fourier series form), then harmonic distortion is present.

If

$$V_o(t) = K[V_1 \cos(\omega t + \theta_1) + V_2 \cos(2\omega t + \theta_2) + V_3 \cos(3\omega t + \theta_3)$$
$$+ \cdots + V_n \cos(n\omega t + \theta_n)]$$

is produced from an input of $V_1 \cos \omega t$, then harmonic distortion is:

$$\% \text{ 2nd H.D.} = \frac{V_2}{V_1} \times 100$$

$$\% \text{ total H.D.} = \frac{\sqrt{V_2^2 + V_3^2 + \cdots + V_n^2}}{V_1} \times 100$$

PROBLEM 6-16. AMPLIFIER CLASS

The common emitter connected transistor amplifier circuit shown below has the following characteristics:

$$h_{FE} = 50$$

$$V_{BE} = 0.6 \text{ volt}$$

$$I_B = 20 \text{ } \mu\text{A}$$

183

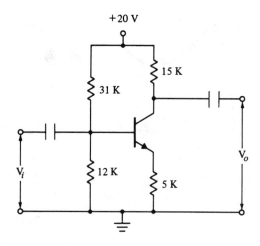

Determine the Q-point, amplifier class, voltage gain, and V_0.

Solution:

$$I_C = h_{FE} I_B = (50)(20 \times 10^{-6}) = 1 \text{ mA}$$

$$V_C = V_{CC} - I_C R_C = 20 - 15 = 5 \text{ volts}$$

$$V_E = I_E R_E \approx I_C R_E = (10^{-3})(5 \times 10^3) = 5 \text{ volts}$$

Since $V_C = V_E = 5$ volts, $V_{CE} = 0$ and the transistor is saturated.
Thus, the Q-point is at $V_{CE} = 0$ and $I_C = 1$ mA.

(answer)   The circuit is operating as a class B amplifier, since the output changes only when the input goes negative, as shown in the following figure.

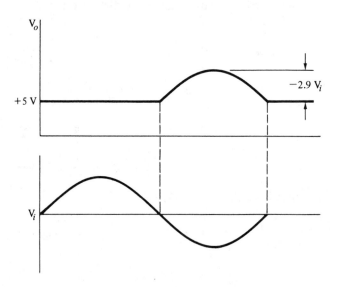

$$A_V = -\frac{h_{FE} R_l}{hie} = \frac{-h_{FE} R_l}{R_b + \dfrac{R_e}{1 - \alpha}} = \frac{V_o}{V_i}$$

$$b_{FE} = 50$$

$$R_b = \frac{(31\text{ K})\,(12\text{ K})}{43\text{ K}} = 8.65\text{ K}\Omega$$

$$R_l = 15\text{ K}\Omega$$

$$R_e = 5\text{ K}\Omega$$

$$\alpha = \frac{\beta}{\beta + 1} = \frac{50}{51} = 0.98$$

$$A_V = \frac{(-50)\,(15 \times 10^3)}{8650 + \dfrac{5000}{0.02}} = -2.9 \qquad \text{(answer)}$$

$$V_o = (A_V)\,(V_i) = -2.9\,V_i \qquad \text{(answer)}$$

If your answers are correct, go on to the next section.
If your answers are not correct, review p. 161.

POWER SUPPLY CIRCUITS

Design of power supplies and power supply circuits is an old and diverse speciality. The following problems illustrate only the simplest type of ac/dc converters.

PROBLEM 6-17.   ZENER REGULATOR

Design a zener diode voltage regulator circuit having the following specifications:

$$V_{\text{out}} = 7.5\text{ VDC}, \quad R_L = 250\ \Omega, \quad V_{\text{in}} = 15\text{ VDC}$$

Define the zener diode characteristics and series resistor.

Solution:

The circuit is as follows:

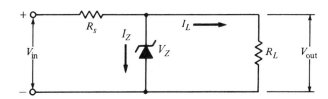

$$V_Z = V_{\text{out}} = 7.5 \text{ volts} \qquad \text{(answer)}$$

$$I_L = \frac{V_Z}{R_L} = \frac{7.5}{250} = 30 \text{ ma}$$

$$P_L = V_Z I_L = 7.5 \times 30 = 225 \text{ mw}$$

For a safety margin, the zener diode should be rated at about three times the maximum load. Thus:

$$P_Z = 3P_L = 675 \text{ mw,}$$

(answer)   so specify the next higher practical value, namely, 1 watt.

(answer)   For safe zener current rating, $I_Z = 3I_L = 90$ ma. Select $I_Z = 100$ ma

$$R_s = \frac{(V_{\text{in max}} - V_Z)^2}{P_Z} = \frac{(15 - 7.5)^2}{0.675} = 83.33 \ \Omega$$

(answer)   To assure safe current limiting, select $R_s = 91 \ \Omega$.

$$P_{R_s} = \frac{(15 - 7.5)^2}{R_s} = \frac{7.5^2}{91} = 0.62 \text{ watt}$$

(answer)   Select a 1 watt resistor for R_s.

To summarize, the zener diode is 7.5 volts, 100 ma, 1 watt.
The series resistor is 91 ohms, 1 watt.

> If your answers are correct, go on to Problem 6-18.
> If your answers are not correct, review Lenk[23], Chapter 6.

PROBLEM 6-18.   RIPPLE FACTOR

A vacuum tube audio amplifier is to be powered from a dc supply. The power supply uses a series choke in a full-wave rectifier circuit as shown below:

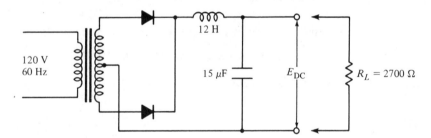

The transformer is 120:480 volt center-tapped, and the drop across each diode is 10 volts. Determine the output voltage and ripple factor.

Solution:

The output of a rectifier circuit without a series choke is equal to the peak of the sine wave (ignoring ripple). However, with a choke in the circuit as shown, the output voltage becomes the average of a sine wave peak. If $X_C \ll R_L$, the capacitor effectively bypasses the ac current around the load. In this case:

$$X_C = \frac{1}{2\omega c} = \frac{1}{4\pi \times 60 \times 15 \times 10^{-6}} = 88.4 \ \Omega$$

(2ω is used because the ripple frequency of a full-wave rectifier is twice the line frequency.)

Thus, it is seen that X_C is, indeed, much less than R_L.

$$E_{\text{peak}} = \sqrt{2} \times 240 = 339.4$$

$$E_{\text{DC}} = \frac{2}{\pi} [E_{\text{peak}} - V_{\text{diode}}] = \frac{2}{\pi} [339.4 - 10] = 209.7 \text{ volts} \qquad \text{(answer)}$$

From page 145 of Ryder[20]:

$$\text{Ripple Factor} = \frac{0.48}{(2\omega)^2 LC - 1} = \frac{0.48}{5.68 \times 10^5 \times 12 \times 15 \times 10^{-6} - 1}$$

$$= 0.464\% \qquad \text{(answer)}$$

If your answers are correct, go on to the next chapter.

If your answers are not correct, review Ryder[20] or an equivalent text on rectifiers.

7 Communications

Discussion of and problems related to low-frequency transmission, RF transmission, and attenuation.

INTRODUCTION

This chapter deals with communications problems that fall into either of two frequency bands: low frequency or radio frequency. The low-frequency band is generally that range between 500 Hz and 1 MHz; this includes telephone transmission problems. The RF range is considered to be anything above 1 MHz; this includes radio and television broadcasting, antennas, and waveguides.

Each band has its own transmission line characteristics, permitting certain unique assumptions to be made.

LOW-FREQUENCY TRANSMISSION

In this section, the transmission line is treated as a circuit having distributed parameters defined below:

R = series resistance, ohms per unit length of line (includes both wires)
L = series inductance, henries per unit length of line
C = capacitance between conductors, farads per unit length of line
G = shunt leakage conductance between conductors, mhos per unit length of line
ωL = series reactance, ohms per unit length of line
$Z = R + j\omega L$ = series impedance, ohms per unit length of line

ωC = shunt susceptance, mhos per unit length of line

$$Z_o = \sqrt{\frac{Z}{Y}} = \sqrt{\frac{j\omega L}{j\omega C}} = \sqrt{\frac{L}{C}} = \sqrt{Z_{oc} Z_{sc}} = \text{characteristic impedance}$$

$Y = G + j\omega C$ = shunt admittance, mhos per unit length of line

s = distance to point of observation, measured from receiving (or sending) end of the line

I = current at any point in the line

E = voltage between the conductors at any point

l = length of the line

γ = propagation constant, per unit length of line

$$\gamma = \sqrt{ZY} = \alpha + j\beta$$

α = attenuation constant, db or neper per unit length of line (1 neper = 8.686 db)

β = phase constant, angle per unit length of line

$E = E_s e^{-\gamma s} = E_s e^{-\alpha s} e^{-j\beta s}$

$I = I_s e^{-\gamma s} = I_s e^{-\alpha s} e^{-j\beta s}$

$e^{-\alpha s}$ = attenuation

$e^{-j\beta s}$ = phase factor

λ = wavelength = $2\pi/\beta = v/f$

$v = \omega/\beta$ = velocity of propagation along the line, miles per second if β is in radians per mile, or meters per second if β is in radians per meter

$\omega = 2\pi f$ in radians

f = frequency in Hz

Velocity v is reduced by $1/\sqrt{\epsilon}$ in a medium having a dielectric constant of ϵ.

PROBLEM 7-1. COMMUNICATION LINE CHARACTERISTICS

An open-wire communication line is 5 miles long and is terminated in its characteristic impedance, Z_o. The line parameters per mile are:

$$R = 75\ \Omega \qquad C = 0.1\ \mu f$$
$$L = 1\ mh \qquad G = 0.05 \times 10^{-6}\ \mho$$

Determine the following at 1 KHz:

a. characteristic impedance Z_o
b. propagation constant γ
c. attenuation α
d. phase shift β
e. velocity v
f. wavelength λ

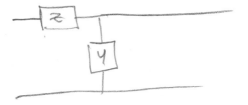

Solution:

a. $Z = R + j\omega L = 75 + j2\pi \times 10^3 \times 10^{-3} = 75 + j6.28 = 75.26\underline{/4.79°}\ \Omega$

$Y = G + j\omega C = 0.05 \times 10^{-6} + j2\pi \times 10^3 \times 10^{-7} = 0.05 \times 10^{-6} + j6.28 \times 10^{-4}$

$$= 6.28 \times 10^{-4} \underline{/90°}\ \mho$$

(answer) $$Z_o = \sqrt{\frac{Z}{Y}} = \sqrt{\frac{75.26\underline{/4.79°}}{6.28 \times 10^{-4}\underline{/90°}}} = 346.18\underline{/-42.61°}\ \Omega$$

(answer) b. $\gamma = \sqrt{ZY} = \sqrt{(75.26\underline{/4.79°})(6.28 \times 10^{-4}\underline{/90°})} = 0.2174\underline{/47.4°} = 0.147 + j0.16$

$\gamma = \alpha + j\beta$

(answer) c. $\alpha = 0.147\ \dfrac{\text{neper}}{\text{mile}} \times \dfrac{8.686\ \text{db}}{\text{neper}} = 1.28\ \text{db/mile}$

(answer) d. $\beta = \dfrac{0.16\ \text{radian}}{\text{mile}}$

(answer) e. $v = \dfrac{\omega}{\beta} = \dfrac{2\pi \times 10^3}{0.16} = 39{,}270\ \text{miles/second}$

(answer) f. $\lambda = \dfrac{2\pi}{\beta} = \dfrac{6.28}{0.16} = 39.27\ \text{miles}$

Also, $\lambda = \dfrac{v}{f} = \dfrac{39{,}270}{10^3} = 39.27\ \text{miles}$

If your answers are correct, continue to the next problem.
If your answers are not correct, review pp. 188 and 189.

PROBLEM 7-2. LF TRANSMISSION LINE POWER

A two-volt generator supplies power to a 150-mile open-wire line terminated in its characteristic impedance and having the following characteristics:

$$R = 10\ \Omega/\text{mile}$$
$$L = 4\ \text{mh/mile}$$
$$G = 1\ \mu\text{mho/mile}$$
$$C = 0.01\ \mu\text{f/mile}$$

Calculate the receiving end power at a frequency of 800 Hz.

Solution:

$$Z = R + j\omega L = 10 + j2\pi \times 800 \times 4 \times 10^{-3} = 10 + j20.11 = 22.46\underline{/63.56}\ \Omega/\text{mile}$$

$$Y = G + j\omega C = 10^{-6} + j2\pi \times 800 \times 10^{-8} = (1 + j50.3)\,10^{-6} = 50.3 \times 10^{-6}\underline{/90°}\ \mho/\text{mile}$$

$$Z_o = \sqrt{\frac{Z}{Y}} = \sqrt{\frac{22.46\underline{/63.56°}}{50.3 \times 10^{-6}\underline{/90°}}} = 668.4\underline{/-13.22°}\ \Omega$$

$$I_S = \frac{E_S}{Z_o} = \frac{2\underline{/0°}}{668.4\underline{/-13.22}} = 3\underline{/13.22°}\ \text{mA}$$

$$\gamma = \sqrt{ZY} = 0.0336\underline{/76.8°}/\text{mile}$$

$$\alpha = 0.0336\cos 76.8° = 0.0077\ \text{neper/mile}$$

190

$$\beta = 0.0336 \sin 76.8° = 0.0327 \text{ rad/mile}$$

$$\frac{I_R}{I_S} = e^{-\gamma l} = e^{-\alpha l}e^{-j\beta l} = e^{-1.16}e^{-j4.91}$$

$$e^{-j4.91} = \underline{/-4.91} \text{ rad} = \underline{/-281°}$$

$$I_R = I_S\, e^{-1.16}\underline{/-281°} = (3\underline{/13.22°})(0.31\underline{/-281°})$$

$$= 0.94\underline{/-268°} \text{ mA}$$

$$E_R = I_R\, Z_o = (0.94 \times 10^{-3}\underline{/-268°})(668.4\underline{/-13°})$$

$$= 0.628\underline{/-281°} \text{ volt}$$

$$P_R = E_R\, I_R\, \cos\theta = 0.628 \times 0.94 \times 10^{-3}\cos 13.22°$$

$$= 0.575 \text{ mW} \qquad\qquad\qquad \text{(answer)}$$

If your answer is correct, go on to Problem 7-3.
If your answer is not correct, review pp. 189 and 190.

PROBLEM 7-3. TRANSMISSION LINE MAXIMUM POWER

A low-frequency transmission line is depicted by the following circuit. Determine the value of the terminating resistor, R_L, in order to develop maximum power in the load. Generator frequency is 1500 Hz.

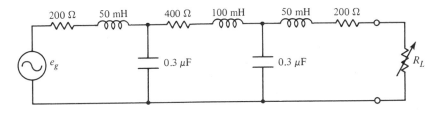

Solution:

$$Z = 400 + j\omega L = 400 + j2\pi \times 1500 \times 100 \times 10^{-3}$$

$$= 400 + j942.5 = 1023.85\underline{/67°} \ \Omega$$

$$Y = G + j\omega C = j2\pi \times 1500 \times 0.6 \times 10^{-6} = 5.66 \times 10^{-3}\underline{/90°} \text{ v}$$

$$Z_o = \sqrt{\frac{Z}{Y}} = \sqrt{\frac{1023.85\underline{/67°}}{5.66 \times 10^{-3}\underline{/90°}}} = 601.5\underline{/-11.5°} \ \Omega$$

Maximum power is achieved when $R_L = |Z_o| = 601.5 \ \Omega$ (answer)

If your answer is correct, go on to the next section.
If your answer is not correct, review Problem 7-1 and maximum power transfer theory corollary on p. 19.

RF TRANSMISSION

In the preceding section, very few simplifying assumptions could be made since, in the low-frequency range, resistances and reactances are of similar orders of magnitude. In the RF range, skin effect and other phenomena unique to high frequencies allow certain simplifying assumptions to be made. These apply both to open-wire and coaxial transmission lines and include the following:

a. Skin effect is significant, so that current may be assumed to flow on conductor surfaces, internal inductance then being zero.
b. $\omega L \gg \omega R$ when computing Z, because resistance increases because of skin effect with $\sqrt{f}$, while reactance increases directly with f.
c. Shunt G may be considered to be zero.

If R is small, the line is said to have *small dissipation*. Small-dissipation lines are not discussed here; however, this is a useful concept when lines are used as circuit elements or where resonance is involved. Reference to a text on the subject will supply information as required.

If R is zero, the line has *zero dissipation (dissipationless line)*. Under these conditions, transmission of power is highly efficient. The dissipationless line is assumed in the following paragraphs and examples.

Two types of transmission lines are used in this section: open-wire and coaxial. Their parameters are discussed in the next two paragraphs.

When discussing the transmission line, the model of a T-section, shown below, is used.

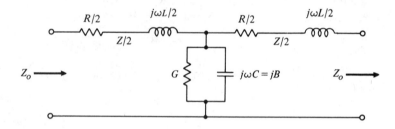

Open-Wire Line Parameters

$$L = \frac{\mu_0}{2\pi} \ln \frac{d}{a} \times 2 \text{ wires} = 4 \times 10^{-7} \ln \frac{d}{a} \text{ henry/meter}$$

where

$$\mu_0 = 4\pi \times 10^{-7} = \text{permeability of air}$$

Generally, magnetic permeability μ is:

$$\mu = \mu_r \mu_v$$

where

μ_r is relative permeability of a particular material
μ_v is magnetic permeability of space = $4\pi \times 10^{-7}$ MKS
μ_0 is magnetic permeability of air = $4\pi \times 10^{-7}$ MKS

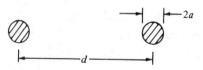

Internal flux and inductance of wires is assumed zero.

$$C = \frac{\pi\epsilon}{\ln\dfrac{d}{a}} \text{ farad/meter} = \frac{27.7}{\ln\dfrac{d}{a}} \text{ pf/m}$$

where

$\epsilon = \epsilon_r\epsilon_v$

ϵ_r is dielectric constant or relative permittivity = 1 for air

ϵ_v is permittivity of space = $10^{-9}/36\pi$ = 8.85×10^{-12} MKS

For round copper conductors,

$$\frac{R_{ac}}{R_{dc}} = \frac{a\sqrt{\pi f \mu \sigma}}{2} = 7.53\, a\sqrt{f}$$

where

$$\mu = 4\pi \times 10^{-7} \text{ (Cu is same as space)}$$

$$\sigma = 5.75 \times 10^7 \text{ } \mho/m \text{ (conductivity of Cu at } 20°C)$$

$$R_{ac} = \frac{8.33 \times 10^{-8} \sqrt{f}}{a} \text{ } \Omega/m \text{ of line for spacing} > 20\, a$$

Coax Line Parameters

$$L = \frac{\mu_v}{2\pi} \ln\frac{b}{a} = 2 \times 10^{-7} \ln\frac{b}{a} \text{ henry/meter}$$

Capacitance is not affected by frequency (except as frequency may alter the relative permittivity of the dielectric):

$$C = \frac{2\pi\epsilon}{\ln\dfrac{b}{a}} = \frac{55.5\, \epsilon_r}{\ln\dfrac{b}{a}} \text{ pf/m}$$

$$R_{ac} = 4.16 \times 10^{-8} \sqrt{f}\left[\frac{1}{a} + \frac{1}{b}\right]$$

where

a is outer radius of inner conductor, in meters

b is inner radius of outer conductor, in meters

Shunt losses of air dielectric lines are zero, but where solid dielectric materials are used, conductance losses sometimes exist, especially at very high frequencies. The quality of the dielectric may be measured in terms of power factor. The shunt susceptance is:

$$Y = G + j\omega C$$

and the power factor is:

$$PF = \frac{G}{\sqrt{G^2 + \omega^2 C^2}}$$

If $G \ll \omega C$,

$$PF = \frac{G}{\omega C} \quad \text{and} \quad G = \omega C \times PF$$

The quality of the dielectric may be expressed in terms of *dissipation factor*, which is the ratio of energy dissipated to energy stored in the dielectric per cycle. For good dielectrics with small PF angles ($G \ll \omega C$), the dissipation factor and *PF* are equal in magnitude.

Zero Dissipation Line Constants

The line parameters for a zero dissipation line are:

$$Z = j\omega L$$

$$Y = j\omega C$$

so that the characteristic impedance Z_0 may be written:

$$Z_0 = \sqrt{\frac{Z}{Y}} = \sqrt{\frac{j\omega L}{j\omega C}} = \sqrt{\frac{L}{C}} \text{ ohms}$$

This value is wholly resistive and may be given the symbol R_0. Propagation constant is:

$$\gamma = \sqrt{ZY} = \sqrt{-\omega^2 LC} = j\omega \sqrt{LC} = \alpha + j\beta$$

where

$$\alpha = 0 \text{ (attenuation constant)}$$

$$\beta = \omega \sqrt{LC} \text{ rad/m (phase constant)}$$

$$= \frac{2\pi f}{v} = \frac{2\pi}{\lambda}$$

Velocity of propagation is:

$$v = \frac{\omega}{\beta} = \frac{1}{\sqrt{LC}} \text{ m/sec}$$

For an *open-wire* line the velocity of propagation is:

$$v = 3 \times 10^8 \text{ m/sec}$$

and

$$R_0 = 120 \ln \frac{d}{a} \, \Omega, \quad \text{for} \quad \frac{d}{a} > 10$$

For a *coax* line:

$$v = \frac{3 \times 10^8}{\sqrt{\epsilon_r}} \text{ m/sec}$$

and

$$R_0 = \frac{60}{\sqrt{\epsilon_r}} \ln \frac{b}{a} \, \Omega$$

where ϵ_r is the dielectric constant between conductors.

194

Zero Dissipation Line Voltages and Currents

In a dissipationless line, attenuation is zero, $Z_0 = R_0$, and voltage and current may be found at any point s units distant from the receiving end of a transmission line from the following equations:

$$E = E_R \cos \frac{2\pi s}{\lambda} + j I_R R_0 \sin \frac{2\pi s}{\lambda}$$

$$I = I_R \cos \frac{2\pi s}{\lambda} + j \frac{E_R}{R_0} \sin \frac{2\pi s}{\lambda}$$

$$Z = E/I$$

where

  E_R is the receiving end voltage
  I_R is the receiving end current
  λ is the wavelength in meters

If the line is open-circuited, $I_R = 0$, and

$$E_{oc} = E_R \cos \frac{2\pi s}{\lambda}$$

$$I_{oc} = \frac{j E_R}{R_0} \sin \frac{2\pi s}{\lambda}$$

The current and voltage are in quadrature everywhere, and no power is transmitted along the line. If the line is short-circuited, $E_R = 0$, and,

$$E_{sc} = j I_R R_0 \sin \frac{2\pi s}{\lambda}$$

$$I_{sc} = I_R \cos \frac{2\pi s}{\lambda}$$

Again, the current and voltage are in quadrature, but the current and voltage waves are shifted $\lambda/4$ from the positions for the open-circuit case. If the line is terminated in a resistance R_R greater than R_0, the *reflection coefficient* K will be positive and the voltage and current conditions on the line will be intermediate to the open circuit and R_0-terminated conditions. K is defined by the formula:

$$K = \frac{Z_R - Z_0}{Z_R + Z_0} \underline{/\phi} = \frac{\text{reflected voltage at load}}{\text{incident voltage at load}}$$

For example, if $R_R = 3R_0$, the value of K is 0.5 and the *incident wave* has an amplitude twice that of the *reflected wave*.

The term "incident wave" refers to a voltage or current wave progressing from the source toward the load. The term "reflected wave" is the voltage or current wave moving from the load toward the source. The magnitude of the reflected wave is dependent on the value of K. The actual voltage at any point on the transmission line is the *vector sum* of the incident and reflected wave voltage at that point. The resultant total voltage wave appears to stand still on the line, oscillating in magnitude with time but having fixed positions of maxima and minima. Such a wave is known as a *standing wave*.

If the line is terminated in $Z_R = R_0$, K and the reflected wave become zero and the voltage on the line is:

$$E = E_R \, e^{j\beta s}$$

where

$\beta = 2\pi/\lambda$, the phase constant in radians/meter
s is distance from the receiving end, in meters

which represents a constant voltage magnitude with continuously varying phase angle along the line. Similarly,

$$I = I_R \, e^{j\beta s}$$

Voltage and current waveforms for different load terminations are shown in Figure 7-1.

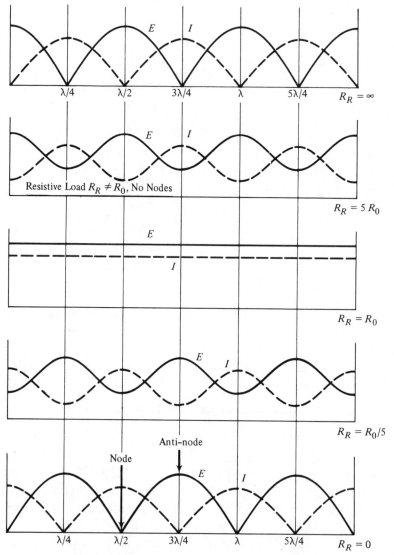

Figure 7-1. Voltages and currents on a dissipationless line.

PROBLEM 7-4. REFLECTION COEFFICIENT

The load end reflection coefficient on a transmission line is given by the following formula:

$$K = \left| \frac{Z_L - Z_o}{Z_L + Z_o} \right| \underline{/\phi^\circ}$$

Determine the ratio of reflected voltage to incident voltage s meters from the load for a zero dissipation transmission line.

Solution:

$$K = \frac{\text{reflected voltage at load}}{\text{incident voltage at load}} = \frac{E_L{}''}{E_L{}'}$$

At s meters from the load end,

$$E_s{}'' = E_L{}'' e^{-(\alpha + j\beta)s}$$

$$E_s' = E_L' e^{(\alpha + j\beta)s}$$

where

   E' is incident voltage
   E'' is reflected voltage
   α is attenuation constant in nepers/meter
   β is phase constant in radians

$$K_s = \frac{E_s{}''}{E_s{}'} = \frac{E_L{}'' e^{-(\alpha + j\beta)s}}{E_L{}' e^{(\alpha + j\beta)s}} = \left| \frac{Z_L - Z_o}{Z_L + Z_o} \right| e^{-2\alpha s} \underline{/\phi - 2\beta s} \qquad \text{(answer)}$$

If your answer is correct, go on to the next section.
If your answer is not correct, review pp. 195 and 196.

Standing Waves

In standing-wave situations, *nodes* are points of zero voltage or current, as occurs in the case of an open-circuit or a short-circuit load. In cases where there is a finite load where $R_R \neq R_0$, there is a standing wave with maximum and minimum points, but not nodes. A line terminated in R_0 has no standing waves and thus no nodes; it is called a *smooth* line. The ratio of maximum to minimum magnitudes of voltage or current is called the *standing wave ratio S*, where

$$S = \left| \frac{E_{max}}{E_{min}} \right| = \left| \frac{I_{max}}{I_{min}} \right| = \frac{1 + |K|}{1 - |K|}$$

or

$$|K| = \frac{S - 1}{S + 1} = \frac{|E_{max}| - |E_{min}|}{|E_{max}| + |E_{min}|}$$

where

$$E_{max} = \frac{E_R(Z_R + Z_0)}{2Z_R}(1 + |K|)$$

$$E_{min} = \frac{E_R(Z_R + Z_0)}{2Z_R}(1 - |K|)$$

S is always expressed as a number $\geqslant 1$.

For the special case of the resistive load:

$$S = \frac{R_R}{R_0} \quad \text{for} \quad R_R > R_0$$

$$S = \frac{R_0}{R_R} \quad \text{for} \quad R_R < R_0$$

Zero-Dissipation-Line Input Impedance

The input impedance of a dissipationless line is:

$$Z_S = \frac{E_S}{I_S} = R_0 \left[\frac{Z_R + jR_0 \tan \beta s}{R_0 + jZ_R \tan \beta s} \right]$$

Another convenient expression for Z_S is:

$$Z_S = R_0 \left[\frac{1 + |K| \underline{/\phi - 2\beta s}}{1 - |K| \underline{/\phi - 2\beta s}} \right]$$

where ϕ is the angle of K.

At values of $s = (\phi/2\beta) + (n\lambda/4)$, the numerator and denominator terms are in phase ($n = 0$, 1, 2, . . .). At these points the input of the line is purely resistive, with min and max values occurring every quarter wavelength.

Maximum input impedance (resistive) at $s = (\phi/2\beta) + (n\lambda/2)$ (phasors coincident) is:

$$R_{max} = R_0 \left[\frac{1 + |K|}{1 - |K|} \right] = SR_0$$

Minimum input impedance (resistive) at $s = (\phi/2\beta) + (2n - 1)\lambda/4$ (phasors coincident) is:

$$R_{min} = R_0 \left[\frac{1 - |K|}{1 + |K|} \right] = \frac{R_0}{S}$$

$$= R_0 \left[\frac{Z_R + jR_0 \tan \frac{2\pi s_2}{\lambda}}{R_0 + jZ_R \tan \frac{2\pi s_2}{\lambda}} \right]$$

where s_2 is distance of the first voltage minimum from the load.

Solving for load impedance yields:

$$Z_R = R_0 \left[\frac{1 - jS \tan \frac{2\pi s_2}{\lambda}}{S - j \tan \frac{2\pi s_2}{\lambda}} \right]$$

The point of voltage minimum is measured rather than voltage maximum because it is usually possible to determine the exact point of the minimum with greater accuracy.

For a short-circuited line, the input impedance is:

$$Z_{sc} = jR_0 \tan \beta s$$

For an open-circuited line, the input impedance is:

$$Z_{oc} = -jR_0 \cot \beta s$$

The characteristic impedance of the line is:

$$Z_0 = \sqrt{Z_{oc} Z_{sc}}$$

Both cases are purely reactive with no power dissipation, alternating between inductance and capacitance each quarter wavelength.

Reflection Losses at Any Point

If a line is not matched to its load, the energy delivered by the line to the load is less than if the impedances were matched. The reflection due to the mismatch results in the following relations:

$$|E_{max}| = |E_i| + |E_r|$$

$$|E_{min}| = |E_i| - |E_r|$$

$$S = \frac{E_{max}}{E_{min}} = \frac{|E_i| + |E_r|}{|E_i| - |E_r|}$$

Power

The expression for power passing along the line and delivered to the load is:

$$P = \frac{|E_{max}| \cdot |E_{min}|}{R_0} = |I_{max}| \cdot |I_{min}| R_0 = \frac{|E_i|^2 - |E_r|^2}{R_0}$$

where

$$E_{max} = \frac{I_R |Z_R + R_0|}{2} (1 + |K|)$$

$$I_{max} = \frac{E_{max}}{R_0}$$

199

$$R_0 = \frac{E_{max}}{I_{max}} = \frac{E_{min}}{I_{min}}$$

$$E_{min} = \frac{I_R |Z_R + R_0|}{2} (1 - |K|)$$

$$I_{min} = \frac{E_{min}}{R_0}$$

$$R_{max} = \frac{E_{max}}{I_{min}} = SR_0$$

$$R_{min} = \frac{E_{min}}{I_{max}} = \frac{R_0}{R}$$

The greatest amount of power is transmitted if $|E_{max}| = |E_{min}|$, or the line is smooth with no standing wave (i.e., $S = 1$) and with R_0 termination.

The ratio of power P delivered to the load to the power P_i transmitted by the incident wave is:

$$\frac{P}{P_i} = \frac{P_i - P_r}{P_i} = \frac{|E_i|^2 - |E_r|^2}{E_i^2} = 1 - |K|^2 = \frac{4S}{(S+1)^2}$$

All of the power is absorbed by the load when $S = 1$.

The Eighth-Wave Line

The input impedance of a line of length $s = \lambda/8$ is:

$$Z_S = R_0 \left[\frac{Z_R + jR_0}{R_0 + jZ_R} \right]$$

If the line is terminated in a pure resistance R_R, then the numerator and denominator have identical magnitudes and,

$$|Z_S| = R_0$$

Thus, an eighth-wave line may be used to transform any resistance to an impedance with a magnitude equal to R_0 of the line.

The Quarter-Wave Line

For $s = \lambda/4$,

$$Z_S = \frac{R_0^2}{Z_R}$$

A quarter-wave section may be used as a transformer to match a load of Z_R to a source of Z_S. A quarter-wave section may be considered to be an impedance inverter, since it can transform

200

a low impedance into a high impedance and vice versa. A $\lambda/4$ short-circuited line looks like an open circuit at the source, and an open-circuited $\lambda/4$ line looks like a short circuit at the source.

The Half-Wave Line

The half-wave line has an input impedance of

$$Z_S = R_0 \left[\frac{Z_R + jR_0 \tan \pi}{R_0 + jZ_R \tan \pi} \right] = Z_R$$

A half-wavelength line may be thought of as a one-to-one transformer. It is useful in connecting a load to a source in cases where the load and source cannot be made adjacent.

Impedance Matching

For greatest efficiency and delivered power, an RF transmission line should be operated as a smooth line with an R_0 termination. If it is not possible to terminate in R_0, then an impedance transforming section added between line and load can be used (discussed on pp. 159 <del>159</del> 207 and 160). Another technique is to use an open or closed stub line of suitable length as a reactance shunted across the line at a designated distance from the load, as shown below:

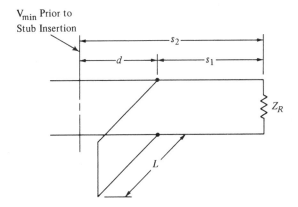

The stub should be located at a distance d measured in either direction from a voltage minimum. Ordinarily, the stub is placed on the load side of the minimum nearest the load. A short-circuited stub is ordinarily preferred to an open-circuited stub because of greater ease in construction and because of the inability to assure high enough insulation resistance for an open circuit. A shorted stub also has a lower loss of energy due to radiation. Dimensions for length and location of the short-circuited stub are given by the formulas below:

$$S = \frac{V_{max}}{V_{min}}$$

$$d = s_2 - s_1 = \left[\text{arc cos} \left(\frac{S - 1}{S + 1} \right) \right] \left(\frac{\lambda}{4\pi} \right)$$

201

$$L = \frac{\lambda}{2\pi} \arc\tan \left(\frac{\sqrt{S}}{S-1} \right)$$

when stub is placed toward the load from V_{min}. When stub is placed toward the source from V_{min},

$$L' = \frac{\lambda}{2} - L$$

PROBLEM 7-5. STUB MATCH

Standing wave measurements made on an RF transmission line with unknown load impedance yield $S = 3.0$. The distance from the load to the nearest minimum = 25 cm = s_2. The distance between the standing wave minima = 40 cm. $Z_0 = 200\ \Omega$.
   Neglecting line dissipation, determine:

a. frequency
b. load parameters
c. ratio of power transmitted to power delivered
d. design and location of stub to match load to line

Solution:

a. Wavelength, $\lambda = 2 \times 40$ cm = 80 cm or 0.8 meter.

(answer)
$$f = \frac{v}{\lambda} = \frac{3 \times 10^8}{0.8} = 375 \text{ MHz}$$

b. $Z_R = Z_0 \left[\dfrac{1 - jS \tan \left(\dfrac{360 s_2}{\lambda} \right)^\circ}{S - j \tan \left(\dfrac{360 s_2}{\lambda} \right)^\circ} \right]$

$$\tan \frac{360^\circ \times 25}{80} = \tan 112.5^\circ = 2.41$$

(answer)
$$Z_R = 200 \left(\frac{1 + j3 \times 2.41}{3 + j2.41} \right) = 379.22\underline{/43.35^\circ}\ \Omega = 276 + j260\ \Omega = R_R + jX_R$$

(answer)
c. $\dfrac{\text{power delivered}}{\text{power transmitted}} = \dfrac{4S}{(S+1)^2} = \dfrac{4 \times 3}{(4)^2} = \dfrac{12}{16} = 0.75$ or 75%

d. $d = \arc\cos \left(\dfrac{S-1}{S+1} \right) \left(\dfrac{\lambda}{4\pi} \right)$

$$d = \arc\cos \left(\frac{3-1}{3+1} \right) \left(\frac{\lambda}{4\pi} \right) = \arc\cos\,(0.5) \left(\frac{80}{4\pi} \right)$$

$$d = \left(60° \frac{\pi}{180°} \text{ rad}\right)(6.37 \text{ cm}) = 1.05 \times 6.37 = 6.67 \text{ cm} \qquad \text{(answer)}$$

$$L = \frac{\lambda}{2\pi} \text{ arc tan}\left(\frac{\sqrt{S}}{S-1}\right) = \frac{80}{2\pi} \text{ arc tan } 0.87 = 12.73 \left(40.89° \frac{\pi}{180°}\right) \qquad \text{(answer)}$$

$$L = (12.73 \text{ cm})(0.717 \text{ rad}) = 9.09 \text{ cm} \qquad \text{(answer)}$$

$$L' = \frac{\lambda}{2} - L = 40 - 9.09 = 30.9 \text{ cm} \qquad \text{(answer)}$$

If your answers are correct, continue to the next section.
If your answers are not correct, review pp. 196 and 198–202.

Smith Chart

A Smith chart is used in many instances to determine the parameters of a lossless transmission line. The Smith chart has the following properties:

1. All possible values of impedance are contained inside the outer circle of unit radius.
2. βs increments are indicated around the outer edge of the chart in terms of wavelengths.
3. A straight edge pivoted at the center and marked in terms of S serves as a distance coordinate to any point on the chart and has the effect of adding constant-S circles to the chart without actually complicating the figure with additional lines.
4. The impedance of a transmission line may be read at any point on the appropriate S-circle.
5. The point at the center of the chart represents the impedance of the line terminated in its characteristic impedance, where $Z/R_0 = 1$ for all distances.
6. The point at the extreme left of the resistance r_a axis represents a short circuit (zero impedance), and the point at the extreme right represents an open circuit (infinite impedance).
7. The outer circle represents $S = \infty$.
8. The chart may be used for admittance as well as impedance, the r_a and x_a axes becoming g_a and b_a axes, with the convention that capacitive susceptance is positive (above) and inductive susceptance is negative (below); the leftmost point is then an open circuit (zero conductance), and the rightmost point is a short circuit (infinite conductance).
9. V_{min} occurs on the real axis. When using impedances, V_{min} occurs on the left half; when using conductances, V_{min} occurs on the right half.

Figure 7-2 is a copy of a Smith chart. The following problems may be solved with its aid.

PROBLEM 7-6. NORMALIZED SENDING IMPEDANCE

Given:

$$Z_R/R_0 = 2.5 + j1.1$$

$$\text{line length} = 30°$$

Find: Normalized Z_s. (*Note:* Smith chart problems require a compass and a straight edge.)

NAME	TITLE	DWG. NO.
SMITH CHART FORM 82-BSPR(9-66)	KAY ELECTRIC COMPANY, PINE BROOK, N.J., © 1966. PRINTED IN U.S.A.	DATE

IMPEDANCE OR ADMITTANCE COORDINATES

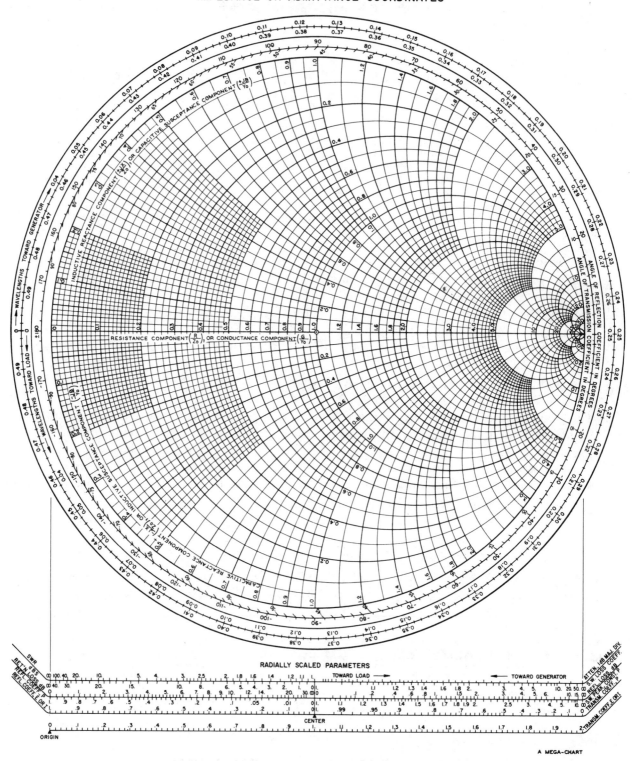

RADIALLY SCALED PARAMETERS

A MEGA-CHART

Figure 7-2. Smith Chart. (Reprinted by permission of Kay Elemetrics Corp.)

Solution:

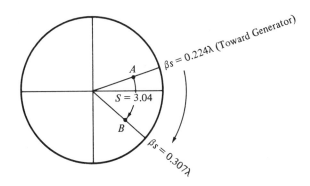

1. Locate load point A at $2.5 + j1.1$ on Smith chart.
2. Draw constant-βs line from origin through A to outer circle reading 0.224λ (toward generator).
3. Calculate line length in terms of λ,

$$\frac{30°}{360°} = 0.083\lambda$$

and move this distance toward generator along outer circle to the point $(0.224 + 0.083)\lambda = 0.307\lambda$.
4. Draw another constant-βs line from origin to 0.307λ on outer circle.
5. Draw a portion of a constant-S circle (center at origin) from point A to point B, where it intersects with the constant βs line drawn in step 4.
6. Read normalized input impedance at point B as $Z_S/R_0 = 1.59 - j1.37$ (capacitive   (answer) reactance).

If your answer is correct, go to the next problem.
If your answer is not correct, review pp. 203 and 204.

PROBLEM 7-7. NORMALIZED Z_R, Z_S, AND K

Given:

$$S = 2.4$$

V_{min} occurs at 0.2λ from load

line length = 0.38λ

Find: Normalized Z_R and Z_S and reflection coefficient, K.

Solution:

V_{min} occurs where S intersects the left half of the resistance axis (least resistance for least voltage).

1. Locate V_{min} as point A on Smith chart.

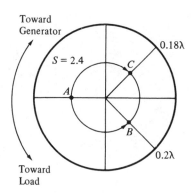

(answer)
(answer)

(answer)

(answer)

2. Move toward load (counter clockwise) to point B where constant βs line $= 0.2\lambda$ intersects with $S = 2.4$ circle.
3. Read $Z_R/R_0 = 1.65 - j0.97$ at point B.
4. Move toward generator (clockwise) from point B to point C a distance of 0.38λ along $S = 2.4$ circle and read $Z_S/R_0 = 1.3 + j1.0$.
5. $|K| = \dfrac{S-1}{S+1} = \dfrac{1.4}{3.4} = 0.412$
6. Read angle of K from βs line at load on the third from outermost scale of chart, giving $-36°$.
   $\therefore K = 0.412\underline{/-36°}$

If your answers are correct, go to the next problem.
If your answers are not correct, review pp. 197, 198, and 203–205.

PROBLEM 7-8.   STUB MATCH, SMITH CHART

Given: $Y_R/G_0 = 2.8 + j1.7$ (capacitive load using admittance form).
   Find: short-circuit stub position and length.

Solution:

1. Locate normalized load admittance at point A on chart $(2.8 + j1.7)$.
2. Drawing a constant-S circle through A shows $S = 3.9$ before use of the stub.

206

3. Locate the $Y/G_0 = 1$ circle on the chart. This is the locus of all points for which the real part of the line conductance is unity, the desired condition at the point of stub connection. Thus, the intersection of the S circle with the $Y/G_0 = 1$ circle at point B is the proper location of the stub toward generator (cw).

4. Since the load is located on the A line ($\beta s = 0.223\lambda$) and the stub is located on the B line ($\beta s = 0.324\lambda$), the stub is located $(0.324 - 0.223)\lambda = 0.101\lambda$ toward the generator from the load.

5. The b_a value at B represents the line susceptance at the stub connection and is -1.5 inductive. The stub must cancel out this imaginary component with a capacitive susceptance of $+1.5$. Find the electrical length of the short-circuited stub by locating the intersection of the $b_a/G_0 = +1.5$ circle on the βs scale ($= 0.157\lambda$) at point C. This intersection occurs $(0.157 + 0.25)\lambda = 0.407\lambda$ from a short circuit at the right end of the real axis or infinite admittance point (measuring toward the load). Thus, a short-circuited stub line 0.407λ in length would have the required capacitive susceptance. An open-circuit stub of 0.157λ would also provide proper match, although open-circuit matching is not as desirable, for reasons given previously.

If your answers are correct, continue to the next section.

If your answers are not correct, review above solution.

Impedance Transformer

There are situations when it is desirable to match a source impedance Z_S to a load impedance Z_R by inserting a quarter-wave (or any odd multiple of a quarter-wave) line in series with the transmission line. Such an application would be to couple a transmission line to the resistive load of an antenna. The quarter-wave matching section must be designed to have a characteristic impedance R_0' so that the antenna resistance R_A is transformed to a value equal to the characteristic impedance R_0 of the transmission line. The line is then terminated in its characteristic impedance and is operated under conditions of no reflection. In this case,

$$R_0' = \sqrt{R_A R_0}$$

This is the value required to achieve critical coupling and maximum power transfer from the transmission line to the load. In general, R_0' of the matching section should equal the geometric mean of the source and load impedances and is equal to:

$$R_0' = \left| \sqrt{Z_S Z_R} \right|$$

ATTENUATION

In many phases of electrical engineering, particularly communications, the use of logarithmic units for power ratios is convenient. If P_1 is the power at one place and P_2 is the power at a second place, they can be related by the formula:

$$\text{bels} = \log \frac{P_1}{P_2}$$

or

$$\text{decibels} = \text{db} = 10 \log \frac{P_1}{P_2}$$

Since $P = V^2/R$,

$$\text{db} = 10 \log \frac{V_1^2}{V_2^2} \times \frac{R_2}{R_1}$$

or

$$\text{db} = 20 \log \frac{V_1}{V_2} + 10 \log \frac{R_1}{R_2}$$

If $R_1 = R_2$,

$$\text{db} = 20 \log \frac{V_1}{V_2}$$

In filter circuits it is sometimes more convenient to use natural logs rather than base-10 (common) logs. In this case the neper, rather than decibel is used. The neper is defined as:

$$N \text{ nepers} = \ln \left| \frac{V_1}{V_2} \right| = \ln \left| \frac{I_1}{I_2} \right|$$

or

$$\left| \frac{V_1}{V_2} \right| = \left| \frac{I_1}{I_2} \right| = e^N$$

Similarly for power ratios:

$$\frac{P_1}{P_2} = e^{2N} = 10^{\#\text{db}/10}$$

Taking the log of both sides yields:

$$1 \text{ neper} = 8.686 \text{ db}$$

A common piece of test equipment in a communication lab is the attenuator. This is a device for attenuating the power entering a given load resistance by a known number of db without changing the total power absorbed by the load resistance and attenuator combined. By use of an attenuator, the power *to the load* is changed, but the power *from the amplifier* is not changed. A simple attenuator circuit and formulas related to this circuit are shown below:

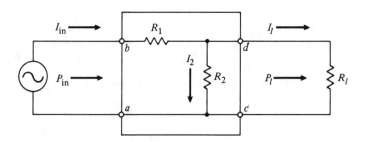

$$R_l = R_1 + \frac{R_2 R_l}{R_2 + R_l}$$

$$R_1 = \frac{R_l^2}{R_2 + R_l}$$

From these formulas it is possible to determine the values for R_1 and R_2 that give the desired power to the load while keeping P_{in} constant and equal to:

$$P_{in} = I_{in}^2 R_{in} = I_{in}^2 R_l$$

If

$$\frac{P_{in}}{P_l} = \frac{I_{in}^2 R_l}{I_l^2 R_l} = \left(\frac{I_{in}}{I_l}\right)^2$$

then

$$db = 20 \log \frac{I_{in}}{I_l}$$

Since R_l and R_2 are in parallel, we can write the voltage between c and d as:

$$I_2 R_2 = I_l R_l, \quad \text{or} \quad I_2 = \left(\frac{I_l R_l}{R_2}\right)$$

Thus,

$$I_{in} = I_l + I_2 = I_l \left(1 + \frac{R_l}{R_2}\right)$$

and

$$db = 20 \log \left(1 + \frac{R_l}{R_2}\right)$$

The equation for R_2 then becomes:

$$R_2 = \frac{R_l}{[10^{db/20} - 1]}$$

PROBLEM 7-9. ATTENUATOR, L-SECTION

Design a fixed attenuator to give a loss of 10 db between the source and a 500 ohm load.

Solution:

$$R_2 = 500 \frac{1}{10^{1/2} - 1} = \frac{500}{2.16} = 231.24 \ \Omega \qquad \text{(answer)}$$

(answer)
$$R_1 = \frac{R_l^2}{R_2 + R_l} = \frac{500^2}{231.24 + 500} = 341.89 \ \Omega$$

If your answers are correct, go on to the next problem.

If your answers are not correct, review pp. 208 and 209.

PROBLEM 7-10.   ATTENUATOR, Π-SECTION

Design a Π-section attenuator for a signal generator feeding a resistive load of 72 ohms. Voltage attenuation shall be 12 db.

Solution:

This problem makes use of a somewhat more efficient attenuator than that of the preceding problem. The Π-section circuit is of the form shown below:

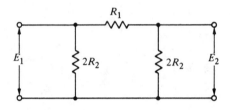

$$db = 20 \log \frac{E_1}{E_2} = 20 \log K,$$

where K is the reflection coefficient.

$$12 = 20 \log K$$

$$K = 3.981$$

$$2R_2 = R_l \left[\frac{K + 1}{K - 1} \right] = 72 \left[\frac{3.981 + 1}{3.981 - 1} \right] = 72 \left[\frac{4.981}{2.981} \right] = 120.3 \ \Omega$$

(answer)
$$R_2 = 60.15 \ \Omega$$

$$R_l = \sqrt{\frac{R_1 R_2}{1 + (R_1 / 4R_2)}}$$

From this,

$$R_1 = \frac{R_l^2}{R_2 - (R_l^2 / 4R_2)}$$

(answer)
$$R_1 = \frac{72^2}{60.15 - 72^2 / 240.6} = 134.3 \ \Omega$$

The Π-section is, thus,

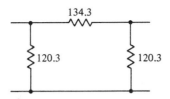

As a point of information, for a T-filter,

$$R_l = \sqrt{R_1 R_2 \left[1 + \frac{R_1}{4R_2}\right]}$$

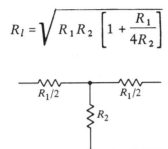

If your answers are correct, go on to the next section.
If your answers are not correct, review the above solution.

ANTENNAS

The following problems are typical of past P.E. exam problems. A text on networks, lines, and fields, such as Ryder,[17] is a good reference for background theory and formulas.
    The following is useful information for working the problems in this section.

1. Impedance of free space = 377 Ω.
2. *Isotropic* means identical in all directions.
3. Radiation resistance of a half-wave $\left(\dfrac{\lambda}{2}\right)$ dipole = 73.26 Ω and is best fed with 72 Ω cable.
4. Gain of a λ/2 dipole = unity (reference) when the dipole is oriented to produce its maximum gain in the same direction as the actual antenna.
5. Surface area of a cylinder = $2\pi rh$.
6. Surface area of a sphere = $4\pi r^2$.
7. Gain of a microwave antenna is given by the formula:

$$G = \frac{4\pi A}{\lambda^2}$$

where

λ = wavelength = $\dfrac{v}{f} = \dfrac{3 \times 10^8}{f}$ meters

A = effective area = ηA_{actual}
η = antenna efficiency

8. Power gain of an actual antenna is the ratio of Poynting vector produced by actual antenna in a particular direction, to value of Poynting vector generated in all directions by an isotropic source of equal power.

9. Poynting vector is: $P_i = \dfrac{W}{4\pi r^2} = \mathcal{E} \times H$ watts/$m^2 = P_{incident}$

10. Path Loss = – Path Gain = $-10 \log\left[\dfrac{P_{incident}}{P_{radiated}}\right] = -10 \log G = 10 \log \dfrac{4\pi r^2}{W} P_r$

PROBLEM 7-11.   VERTICAL ANTENNA

A newly installed vertical antenna at an AM broadcast station transmitter facility has an effective half-power radiation cross section as shown below:

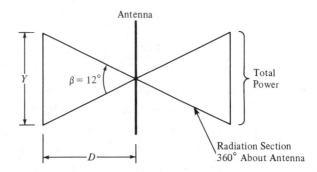

At 80 meters from the antenna, a field strength meter indicates the maximum rms value of field strength is 30 volts/meter.   Calculate the total value of time-average power that the antenna radiates under these conditions.

Solution:

At 80 meters, $E_{\text{rms max}} = 30$ v/m

At half-power points, $E_{\text{rms}} = \left[\dfrac{30}{\sqrt{2}}\right]$ v/m = E

$P = \dfrac{E^2}{377} = \left[\dfrac{30}{\sqrt{2}}\right]^2 \cdot \dfrac{1}{377} = 1.19$ watts/m^2

where 377 Ω is the impedance of free space

$\tan \dfrac{\beta}{2} = 0.5 \dfrac{Y}{80}$, $Y = 160 \tan 6° = 16.82$ Meters

Total area of the cylinder of revolution = $2\pi DY = 2\pi 80 \times 16.82$
$$= 8453 \ m^2$$

(answer) $P_{\text{Total}} = 8453 \times 1.19 = 10,059$ watts

If your answer is correct, go on to Problem 7-12.
If your answer is not correct, review related material in Ryder[17] or a similar text.

PROBLEM 7-12. MICROWAVE ANTENNA

Two microwave stations operating at 5 GHz are 50 kilometers apart. Each has an antenna whose gain is 50 dB greater than isotropic. If 6 watts is applied to the input of the transmitting antenna, what is the signal level at the output terminal of the receiving antenna under free-space conditions? What is the path loss?

Solution:

$$G = 50 \text{ dB} = 10^5, f = 5 \times 10^9 \text{ Hz}, P_T = 6 \text{ watts}$$

$$\lambda = \frac{v}{f} = \frac{3 \times 10^8}{5 \times 10^9} = 6 \times 10^{-2} \text{ m}$$

$$A = \frac{\lambda^2 G}{4\pi} = \frac{(6 \times 10^{-2})^2 10^5}{4\pi} = 28.65 \text{ m}^2 \text{ effective antenna area}$$

$$\text{watts/m}^2 \text{ @ 50 KM} = \frac{P_T G}{\text{surface area of sphere}} = \frac{6 \times 10^5}{4\pi(50 \times 10^3)^2}$$

$$= 1.91 \times 10^{-5} \text{ watts/m}^2$$

$$P_{\text{RCVR}} = A [\text{w/m}^2 \text{ @ 50 KM}] = 28.65 \times 1.91 \times 10^{-5} = 5.47 \times 10^{-4} \text{ watt} \qquad \text{(answer)}$$

$$\text{Path Loss} = -10 \log G = -10 \log \frac{4\pi}{\lambda^2} (4\pi r^2) = -10 \log \frac{4\pi r}{\lambda}^2$$

$$= -20 \log \frac{4\pi \times 50 \times 10^3}{6 \times 10^{-2}} = -20 \log 1.05 \times 10^7$$

$$= -140 \log 1.05 = 140.4 \text{ dB} \qquad \text{(answer)}$$

If your answers are correct, go on to Problem 7-13.
If your answers are not correct, review the introductory material for this section.

PROBLEM 7-13. TWO HALF-WAVE ANTENNAS

A phased array consists of two half-wave antennas located a half-wave apart, as shown in the figure below:

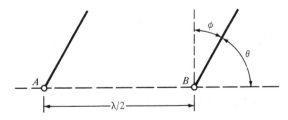

Antenna currents, I_A and I_B, are equal, and I_A leads I_B by 90°. Maximum field strength of each individually excited antenna is 250 mv/m at a distance of 40 KM. Radiation resistance of each antenna is 73.1 Ω (the same as a half-wave dipole).

213

Determine the angle, θ, at which the resultant field strength is maximum, and calculate the field strength of the array for $\theta = 90°$ at a distance of 25 KM.

Solution:

The rms value of the resultant electric field is given by the formula:

$$\mathcal{E}_r = \mathcal{E}_{rms} \cos \left(\pi n \sin \phi + \frac{\delta}{2} \right)$$

where

δ is the phase angle between I_A and I_B, and is positive when I_A leads I_B
n is the distance between A and B in wavelengths, and is usually fractional
Angle ϕ is measured from the normal to the line of the antennas

Thus,

$$\mathcal{E}_r = \mathcal{E}_{rms} \cos \left(\frac{\pi}{2} \sin \phi + \frac{\pi}{4} \right)$$

For maximum $\mathcal{E}_r$,

$$\cos \left(\frac{\pi}{2} \sin \phi + \frac{\pi}{4} \right) = 1$$

or

$$\text{or} \left(\frac{\pi}{2} \sin \phi + \frac{\pi}{4} \right) = 0$$

$$\sin \phi = 0.5, \ \phi = 30°$$

(answer)
$$\therefore \ \theta = 60°$$

At $\theta = 90°$,

$$\cos \left(\frac{\pi}{2} \sin 0° + \frac{\pi}{4} \right) = \cos \frac{\pi}{4} = 0.707$$

The field strength of the array at 25 KM distance is:

(answer)
$$\mathcal{E}_i = 0.707 \times 250 \ \frac{mv}{m} \left[\frac{40}{25} \right]^2 = 282.8 \ mv/m$$

If your answers are correct, go on to Problem 7-14.
If your answers are not correct, review the introductory material for this section.

PROBLEM 7-14. WAVEGUIDE POWER RATING

A standard brass (Cu–Zn) waveguide has the following outer dimensions and wall thickness (inches): 6.66 × 3.41 × 0.08. Assuming the breakdown strength of air is at least 15,000 volts/cm, what is the theoretical maximum power rating for the lowest operating frequency?

Solution:

The sixth edition of *Reference Data for Radio Engineers*[36] page 25-8, lists the characteristics of this waveguide. The last column shows the maximum power rating to be 11.9 megawatts for the lowest operating frequency of 1.12 GHz. (answer)

Note: This type of problem found in the PE Exam does not require calculations. Rather, it requires that the examinee have some background in the field so that he knows there is a handbook available containing the proper information. If the examinee has such a handbook with him, he can solve the problem in less than five minutes.

If your answer is correct, go on to the next chapter.
If your answer is not correct, review the above solution.

8 Logic

Discussions of and problems related to number systems, truth functions, Boolean algebra, switching devices, minimization of Boolean functions, codes and special realizations, and sequential circuits.

INTRODUCTION

The PE Examinations are now beginning to include digital logic. Questions in digital logic can be expected to be included in future PE Exams.

This chapter presents general principles with which you should be familiar. If you would like further review of digital logic, *Introduction to Switching Theory and Logic Design* by Hill and Peterson[26] is a good reference text.

A basic review of the field should include the following subjects:

Number Systems
    Bases other than 10
    Conversion between bases
    Negative numbers

Truth Functions
    Logical reasoning
    Binary connectives
    Physical realizations (AND, OR, NOT; inclusive/exclusive OR)

Boolean Algebra
    Huntington's postulates, duality
    Truth calculus vs. Boolean algebra (+, X, ·)
    DeMorgan's theorem

Venn diagram
Boolean simplification

Switching devices
  Gates
  Flip-flops

Minimization of Boolean functions
  Terms (literal, product, sum, normal, conjunction, disjunction)
  Minterms and maxterms, prime implicants
  Karnaugh map (graphical method of minimization)
  Quine–McClusky method (tabular method of minimization)

Codes and Special Realizations
  Error detection and correction
  Parity check
  Adder (half, full)
  BCD
  Gray (reflected)
  Excess-3
  Hamming
  Distance

Sequential Circuits
  Shift registers
  Counters
  Clocked circuits, sequencers
  Timing considerations
  State diagrams
  Mealy–Moore translations and circuits
  Cycles, races, hazards

DEFINITION OF TERMS, POSTULATES, PROPERTIES

The following terms are used in the field of logic design:

Connectives

Term	Symbols
AND	$\wedge, \cdot, \cap$
(Inclusive) OR	$\vee, +, \cup$
Exclusive OR	$\oplus, \forall$
NOT	$^{-}, \sim, ', *$
NOT AND (NAND)	$\uparrow$ {Sheffer's
NOT OR (NOR)	$\downarrow$ {Strokes }
If A is true, then B is true	$\supset$

Truth Table of Binary Connectives

Connective		∧	·	A		B	⊕	∨	↓	≡	$\bar{B}$		$\bar{A}$	⊃	↑	
A	B	0	1	2	3	4	5	6	7	8	9	10	11	12	13	14	15
0	0	0	0	0	0	0	0	0	0	1	1	1	1	1	1	1	1
0	1	0	0	0	0	1	1	1	1	0	0	0	0	1	1	1	1
1	0	0	0	1	1	0	0	1	1	0	0	1	1	0	0	1	1
1	1	0	1	0	1	0	1	0	1	0	1	0	1	0	1	0	1

Boolean Functions: Terms and Synonyms

Term	Definition	Synonym
Literal	A variable or its complement $(A, \bar{A}, B, \bar{B})$	
Product term	A series of literals related by AND $(A\bar{B}D)$	Conjunction
Sum term	A series of literals related by OR $(\bar{A} + B + \bar{D})$	Disjunction
Normal term	A product or sum term in which no variable appears more than once	

Huntington's Postulates

   I. There exists a set of K objects or elements, subject to an equivalence relation, denoted "=", which satisfies the principle of substitution. By substitution is meant: if $a = b$, then a may be substituted for b in any expression involving b without affecting the validity of the expression.

 IIa. A rule of combination "+" is defined such that $a + b$ is in K whenever a or b are in K.

 IIb. A rule of combination "·" is defined such that $a \cdot b$ (abbreviated ab) is in K whenever both a and b are in K.

IIIa. There exists an element 0 in K such that, for every a in K, $a + 0 = a$.

IIIb. There exists an element 1 in K such that, for every a in K, $a \cdot 1 = a$.

IVa. $a + b = b + a$ ⎫
IVb. $a \cdot b = b \cdot a$ ⎬ cummutative laws.

 Va. $a + (b \cdot c) = (a + b) \cdot (a + c)$ ⎫
 Vb. $a \cdot (b + c) = (a \cdot b) + (a \cdot c)$ ⎬ distributive laws.

 VI. For every element a in K there exists an element a such that

$$a \cdot \bar{a} = 0$$

and

$$a + \bar{a} = 1$$

VII. There are at least two elements X and Y in K such that $X \neq Y$.

218

Duality

Every theorem which can be proved for Boolean algebra has a dual which is also true, as shown below:

$$a + (b \cdot c) = (a + b) \cdot (a + c)$$
$$a \cdot (b + c) = (a \cdot b) + (a \cdot c)$$

DeMorgan's Law (Theorem)

For every pair of elements a and b in K,

$$(\overline{A \cdot B}) = \overline{A} + \overline{B}$$
$$(\overline{A + B}) = \overline{A} \cdot \overline{B}$$

Associative Laws

For any three elements, a, b, and c in K,

$$a + (b + c) = (a + b) + c$$

and

$$a \cdot (b \cdot c) = (a \cdot b) \cdot c$$

Venn Diagrams Using Sets A and B

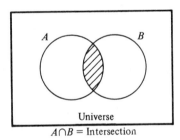

$A \cap B$ = Intersection

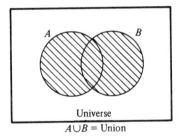

$A \cup B$ = Union

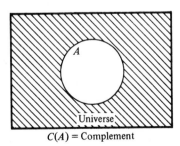

$C(A)$ = Complement

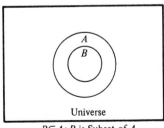

$B \subset A$; B is Subset of A

If a set A contains all objects of the universe, it is called a universal set, A_U. If a set A contains no objects of the universe, it is called a null set A_Z.

Prime Implicant

A *prime implicant* is any sphere of a function which is not totally contained in some larger sphere of the function.

Maxterm (Canonic Sum, Standard Sum)

If a sum contains as many literals as there are variables in a function, then it is called a maxterm.

Combinational

If at any particular time the present value of the outputs is determined solely by the present value of the inputs, such a system is combinational.

Sequential

If the present value of the outputs is dependent not only on the present value of the inputs, but also on the past history of the system, then such a system is sequential.

Gates

Combinational logic is made up of groups of AND and OR gates, with their many variations. These have been implemented over the years as diode logic, resistor–transistor logic (RTL), diode–transistor logic (DTL), direct-coupled transistor logic (DCTL), transistor–transistor logic (TTL), emitter-coupled logic (ECL), and others. There is negative logic ($V_- = 1$, $V_+ = 0$) and positive logic ($V_- = 0$, $V_+ = 1$); the voltage swing away from ground can be either positive or negative. Transistors used can be either PNP or NPN. No matter how gates are implemented hardware-wise, their inputs and outputs can assume only one of two states at any one time. Circuits for various gate hardware mechanizations are shown below.

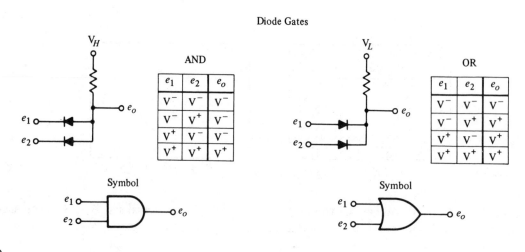

Diode Gates

AND

e_1	e_2	e_o
V^-	V^-	V^-
V^-	V^+	V^-
V^+	V^-	V^-
V^+	V^+	V^+

OR

e_1	e_2	e_o
V^-	V^-	V^-
V^-	V^+	V^+
V^+	V^-	V^+
V^+	V^+	V^+

Symbol

Symbol

DTL Gate

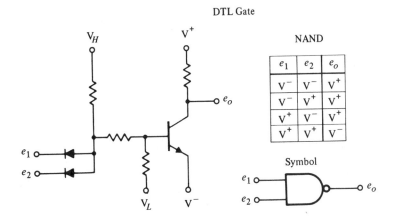

NAND

e_1	e_2	e_o
V^-	V^-	V^+
V^-	V^+	V^+
V^+	V^-	V^+
V^+	V^+	V^-

Symbol

Typical RTL Gate

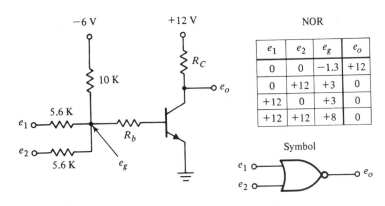

NOR

e_1	e_2	e_g	e_o
0	0	−1.3	+12
0	+12	+3	0
+12	0	+3	0
+12	+12	+8	0

Symbol

TTL Gates (7400 Series)

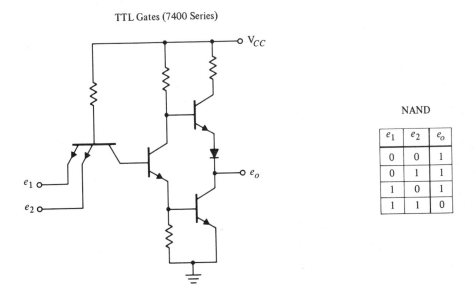

NAND

e_1	e_2	e_o
0	0	1
0	1	1
1	0	1
1	1	0

221

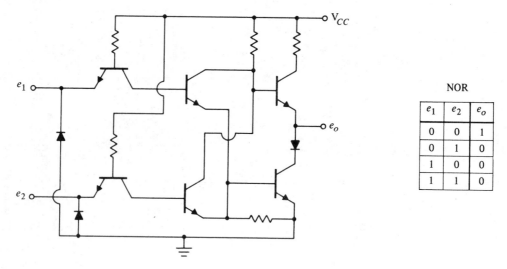

NOR		
e_1	e_2	e_o
0	0	1
0	1	0
1	0	0
1	1	0

Flip-Flops

The ability to store information is an important characteristic of a digital system. The most common hardware storage device is the flip-flop, or bistable multivibrator. The following are various types of flip-flops that are implemented in hardware:

a. *D.* A flip-flop whose output is a function of the input which appeared just prior to the clock pulse.

b. *J–K.* A flip-flop having two inputs, *J* and *K*. At the application of a clock pulse, 1 on the *J* input will set the output to 1; 1 on the *K* input will reset the output to 0; 1 on both inputs will cause the output to change state (toggle); 0 on both inputs results in no change in the output state.

c. *R–S.* A flip-flop having two inputs, *R* and *S*. Operation is the same as the *J–K* flip-flop except that 1 on both inputs is illegal.

d. *R–S–T.* A flip-flop having three inputs, *R*, *S*, and *T*. The *R* and *S* inputs produce outputs as described for the *R–S* flip-flop; the *T* input causes the flip-flop to toggle.

e. *T.* A flip-flop having only one input. A pulse appearing on the input causes it to toggle.

Truth tables for basic flip-flops are shown below:

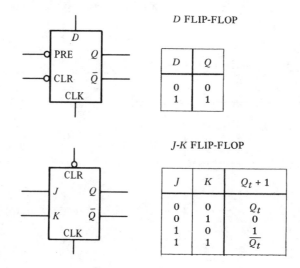

D FLIP-FLOP

D	Q
0	0
1	1

J-K FLIP-FLOP

J	K	$Q_t + 1$
0	0	Q_t
0	1	0
1	0	1
1	1	$\overline{Q_t}$

R-S FLIP-FLOP (LOW ACTIVE INPUTS)

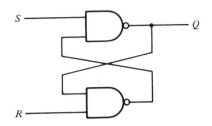

S	R	$Q_t + 1$
0	0	illegal
0	1	1
1	0	0
1	1	Q_t

R-S FLIP-FLOP (HIGH ACTIVE INPUTS)

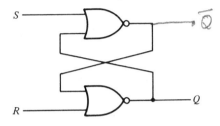

S	R	$Q_t + 1$
0	0	$\overline{Q_t}$
0	1	1
1	0	0
1	1	illegal

PROBLEM 8-1. DeMORGAN'S THEOREM

Using DeMorgan's Theorem, show that the gate pairs shown below are identical.

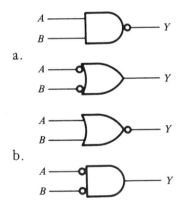

a.

b.

Solution:

a. The first gate is a two-input NAND (e.g., 7400); the second gate is the same part used to perform an OR function. Since they are identical parts, the outputs must be the same if the inputs are the same.

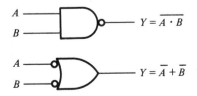

By DeMorgan's Theorem, $\overline{A \cdot B} = \overline{A} + \overline{B}$.   (answer)

b. The second gate pair is a two-input NOR (e.g., 7402) used in two different ways.

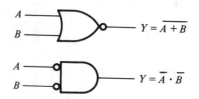

(answer)   By DeMorgan's Theorem, $\overline{A + B} = \overline{A} \cdot \overline{B}$.

If your answers are correct, go to the next problem.
If your answers are not correct, review p. 219.

PROBLEM 8-2.   LOGIC SIMPLIFICATION, KARNAUGH MAP

Simplify the following equation and show the final logic diagram:

$$Y = \overline{A}\,\overline{B}\,\overline{C} + \overline{A}\,B\,\overline{C} + A\,B\,\overline{C} + A\,\overline{B}\,\overline{C} + \overline{A}\,\overline{B}\,C + A\,\overline{B}\,C$$

Solution:

Use a Karnaugh map to assist in simplification:

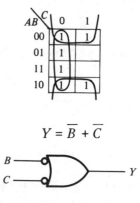

$$Y = \overline{B} + \overline{C}$$

(answer)

If your answer is correct, go to the next problem.
If your answer is not correct, review the above solution.

PROBLEM 8-3.   TIMING DIAGRAM, TRUTH TABLE

For a two-input NAND gate, if the A and B inputs are as shown in the following timing diagram, what is the waveform for output Y? Show the truth table.

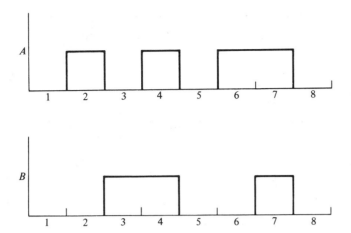

Solution:

For each interval, Y is as shown below:

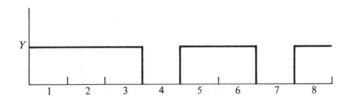

The truth table is:

	A	B	Y
1	0	0	1
2	1	0	1
3	0	1	1
4	1	1	0
5	0	0	1
6	1	0	1
7	1	1	0
8	0	0	1

If your answers are correct, go to the next problem.
If your answers are not correct, review the above solution.

PROBLEM 8-4. LOGIC SEQUENCER INPUT LOGIC

Figure 8-1 shows the block diagram of a sequencer that implements the control loop of Figure 8-2.

External switches S_1 and S_2 determine jump conditions (i.e., when S_1 is ON the sequencer steps from state B to C, and when S_1 is OFF the sequencer steps from state B to E). Use three

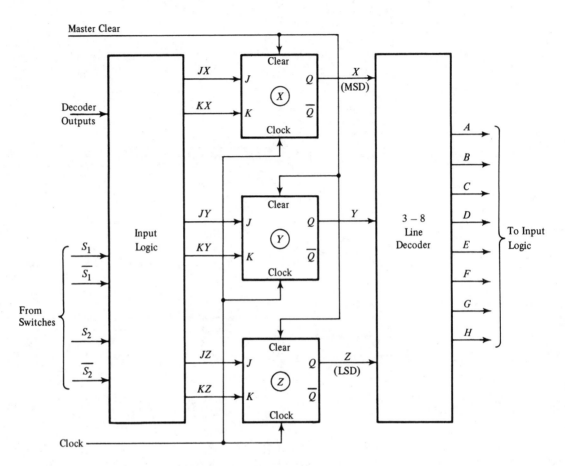

Figure 8-1. Logic sequencer block diagram.

J–K flip-flops whose decoded outputs represent the eight states. The flip-flops are clocked simultaneously by a synchronous clock. Assume the two switches S_1 and S_2 remain unchanged during the active clock pulse time. Start with state $A = \overline{X} \cdot \overline{Y} \cdot \overline{Z}$ as shown in Figure 8-2 (the sequencer unconditionally jumps to this state when the system is master cleared).

a. Assign an X, Y, Z control flip-flop combination to each state by filling in the Karnaugh map below. Only one control flip-flop at a time may toggle when the sequencer is clocked from one state to the next; thus, states joined by lines in Figure 8-2 are adjacent to each other on the Karnaugh map. States A and B are already assigned.

X \ YZ	00	01	11	10
0	A	B		
1				

Karnaugh Map

226

Now transfer these state combinations to Figure 8-2 in the same manner as was done for states *A* and *B*.

b. Write below the five remaining input equations for the three *J–K* control flip-flops as a

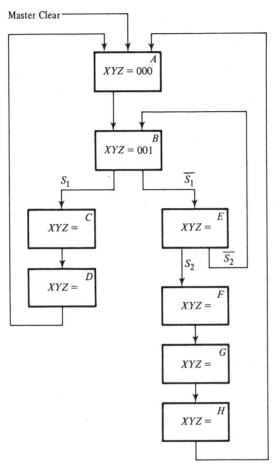

Figure 8-2. Control loop.

function of the eight decoded states and the two switch conditions (the equation for the *J*-input to flip-flop *Z* is shown completed).

$$JZ = A$$
$$KZ =$$
$$JY =$$
$$KY =$$
$$JX =$$
$$KX =$$

Solution:

a. There are two acceptable Karnaugh maps for this problem. One is shown below:

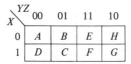

X \ YZ	00	01	11	10
0	A	B	E	H
1	D	C	F	G

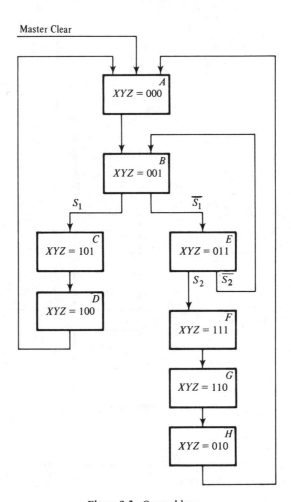

Master Clear

Figure 8-3. Control loop.

The binary code for each state is then entered in the appropriate block of the control loop, as shown. Thus, it is seen that only one flip-flop changes state between adjacent blocks of the control loop. (See Figure 8-3.)

b. The five remaining input equations for the flip-flops are shown below. A J input is set to a logic ONE at the block just prior to the flip-flop switching from a ZERO to a ONE.

The K input is set to a logic ONE at the block just prior to the flip-flop switching from a ONE to a ZERO. These inputs are set up and stable at the time of the next clock pulse.

$$KZ = C + F \qquad \text{(answers)}$$

$$JY = B \cdot \overline{S_1}$$

$$KY = H + E \cdot \overline{S_2}$$

$$JX = B \cdot S_1 + E \cdot S_2$$

$$KX = D + G$$

An alternative solution follows:

X \ YZ	00	01	11	10
0	A	B	C	D
1	H	E	F	G

$$KZ = C + F \qquad \text{(answers)}$$

$$JY = B \cdot S_1 + E \cdot S_2$$

$$KY = D + G$$

$$JX = B \cdot \overline{S_1}$$

$$KX = H + E \cdot \overline{S_2}$$

If your answers are correct, go on to Problem 8-5.
If your answers are not correct, review the above solution and reference 48.

PROBLEM 8-5. UNDERLAPPED TWO-PHASE CLOCK CIRCUIT

The J-K flip-flop circuit shown below is commonly used to generate an underlapped two-phase clock. The flip-flop is the master-slave 74107 in which the output switches on the falling edge of the input clock.

Draw a timing diagram showing the relation of the input and output clocks. Also show another method of developing a two-phase clock using a 7474 edge-triggered D flip-flop.

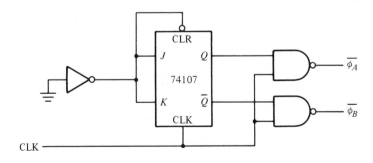

Solution:

The timing diagram for the circuit above is as follows:

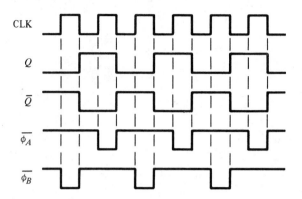

Using a D flip-flop, an alternate circuit with its timing diagram is shown below:

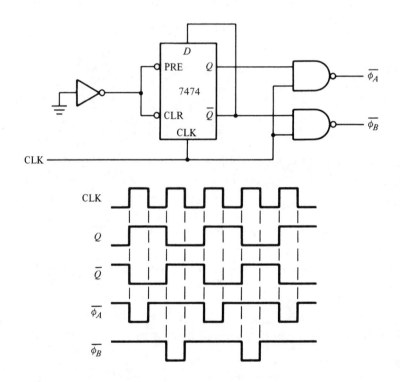

If your solution is correct, go on to Problem 8-6.
If your solution is not correct, refer to pp. 222–223.

PROBLEM 8-6. MULTIPHASE CLOCK CIRCUIT, JOHNSON COUNTER

The Johnson counter is a popular multiphase clock circuit that is used to generate an even number of non-overlapped clock phases using a shift register with inverted output fed back to the input, as shown in the circuit below.

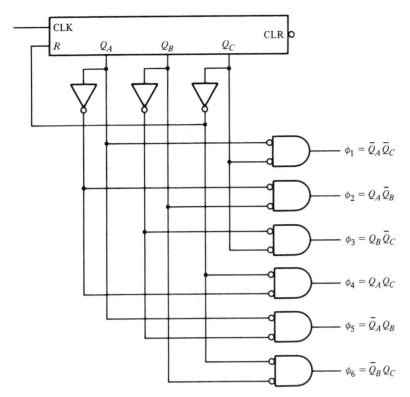

TRUTH TABLE

Q_A	Q_B	Q_C	ϕ
0	0	0	1
1	0	0	2
1	1	0	3
1	1	1	4
0	1	1	5
0	0	1	6

$\phi_1 = \bar{Q}_A \bar{Q}_C$

$\phi_2 = Q_A \bar{Q}_B$

$\phi_3 = Q_B \bar{Q}_C$

$\phi_4 = Q_A Q_C$

$\phi_5 = \bar{Q}_A Q_B$

$\phi_6 = \bar{Q}_B Q_C$

Draw the timing diagram for the six-phase clock circuit showing the six clock phases, the three shift register outputs, and the input clock.

Solution:

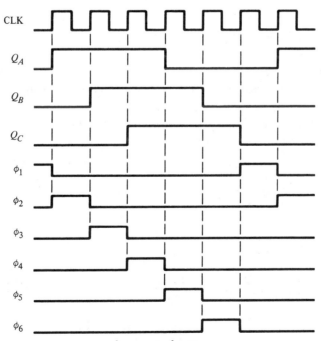

If your solution is correct, go on to the next chapter.

If your solution is not correct, review the preceding problem regarding timing diagrams, and Problem 8-1 on De Morgan's Theorem.

9 Economics

Discussions of and problems related to interest, depreciation, equivalent costs, breakeven analysis, probability, and marginal cost analysis.

INTRODUCTION

Engineering economics is a specialized subject not normally included in the college engineering curriculum. Engineering economics is *not* theoretical, and it is *not* concerned with balance sheets or profit–loss statements. Rather it is a pragmatic study of costs which involve decision-making (often times among investment alternatives) regarding future income and expenses. Engineering economics involves planning and good business judgment.

This chapter reviews the basic concepts of interest, depreciation, equivalent costs, breakeven analysis, probability, and marginal cost analysis. Most PE Examination problems in economics involve one or more of these concepts. Therefore, understanding the following material should prepare the candidate for all but the most esoteric economic problems in the PE Examination.

INTEREST

Definitions of Terms

An understanding of the following terms is necessary when considering the subject of interest:

Interest: Cost of money.

Interest rate: The ratio of the amount (excluding principal) paid at the end of a period to the amount of money owed at the beginning of the period. Sometimes called rate of return, return on investment, and profitability index.

Interest period: The period for which the interest is applied.

Principal: The amount of money owed at the beginning of an interest period.

Simple Interest: Interest paid only on the original principal, not on the interest accrued.

Compound Interest: Interest paid on the accumulated unpaid interest as well as on the original principal.

Nominal Interest Rate: An annual rate expressed as a product of the rate for a shorter period times the number of these periods in a year (e.g., a 1.5% interest rate for one month would yield an 18% nominal interest rate).

Effective Interest Rate: The actual annual interest rate derived from the rate for a shorter period by the following formula:

$$i = \left[1 + \frac{j}{m}\right]^m - 1$$

where

i is the effective annual interest rate,
j is the nominal annual interest rate, and
m is the number of periods in the year.

This amounts to interest compounded on a basis more often than annually, as is the case with savings accounts. For example, money deposited at 6% compounded monthly yields an effective annual interest rate of 6.168%.

As the number of periods m approaches ∞, the effective interest i becomes:

$$i = e^j - 1$$

This is known as *continuous compounding.*

One derivation of this formula follows.
Using the relation, $x = e^{\ln x}$, and Lhopital's theorem,

$$\lim_{m \to \infty} \left[1 + \frac{j}{m}\right]^m = \lim_{m \to \infty} e^{m[\ln(1+j/m)]}$$

$$= \exp\left\{\lim_{m \to \infty}\left[\frac{\ln(1+j/m)}{1/m}\right]\right\}$$

$$= \exp\left[\lim_{m \to \infty}\left\{\frac{\left[\frac{d}{dm}[\ln(1+j/m)]\right]}{\frac{d}{dm}(1/m)}\right\}\right]$$

$$= \exp\left[\lim_{m \to \infty}\left\{\frac{\left[\frac{1}{1+j/m}\right]\left[\frac{-j}{m^2}\right]}{-1/m^2}\right\}\right]$$

$$= \exp\left\{\lim_{m \to \infty}\left[\frac{j}{1+j/m}\right]\right\} = e^j$$

Discount Rate: The actual interest rate when the amount of interest at a stated rate is deducted from the principal at the beginning of the period. For example, if a sum of money P is borrowed for one year and the lender deducts the interest value v immediately, then the effective interest rate i is given by the formula:

$$i = \frac{100 \, v\%}{(P - v)}$$

For example, if a person borrows $1000 *discounted* for one year at 10%, then the effective annual interest rate is

$$i = \frac{100(0.1 \times 1000)}{1000 - (0.1 \times 1000)} = \frac{100 \times 100}{1000 - 100} = \frac{10,000}{900} = 11.1\%$$

Interest Formulas

Economics problems involving time value of money (interest and annuity) require the establishment of relationships among:

P—present value;
F—future value, or sum, at the end of a number of periods;
A—uniform end-of-period payment;
n—number of interest periods; and
i—interest rate per interest period.

There are two basic interest formulas (compound interest and annuity) from which four other formulas are obtained. Some published interest tables contain all six. Most tables contain at least the compound interest formula and two annuity formulas. The six formulas are related as follows:

Compound Interest $= (1 + i)^n$

$$\text{Present Value} = \frac{1}{\text{Compound Interest}}$$

$$\text{Present Value of Annuity} = \left[\frac{(1 + i)^n - 1}{i(1 + i)^n} \right]$$

Future Value of Annuity = (Present Value of Annuity) (Compound Interest)

$$\text{Sinking Fund Factor} = \frac{1}{\text{Future Value of Annuity}}$$

$$\text{Capital Recovery Factor} = \frac{1}{\text{Present Value of Annuity}}$$

Four of these relationships are shown in Table 9-1 together with the name of the formula as found in standard interest tables. These names tend to vary, but the formulas are always the same; when in doubt choose the table having the proper formula.

234

TABLE 9-1. Summary of Interest Formulas.

PAYMENT		VALUE			
			Amount (compounding each period)		
			Formula		Name of Table Used
When	Amount	When	Exact	Functional	
Now	P	n periods hence	$F = P(1 + i)^n$	$F = P\,[CI]_i^n$	compound interest or compound amount factor
n periods hence	F	Now	$P = F(1 + i)^{-n}$	$P = F\,[PV]_i^n$	present value
At end of each period	A	n periods hence	$F = A\left[\dfrac{(1 + i)^n - 1}{i}\right]$	$F = A\,[FVA]_i^n$	future value of annuity or amount of annuity or sinking fund amount factor
At end of each period	A	Now	$P = A\left[\dfrac{(1 + i)^n - 1}{i(1 + i)^n}\right]$ or $P = A\left[\dfrac{1 - (1 + i)^{-n}}{i}\right]$	$P = A\,[PVA]_i^n$	present value of annuity or annuity fund factor

Use of Interest Formulas

The following examples illustrate use of the interest formulas. Rather than calculate the values using appropriate formulas, interest tables may be used for looking up the proper numbers. However, if a table does not contain the desired number, then, for accuracy, the formula must be used. Interpolation should be used with care since the formulas are not linear.

In working the problems it is helpful to draw a time-payment chart of the type shown below:

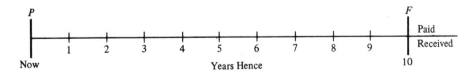

As a point of information, the future value of annuity formula may be derived as follows:

$$F = A + A(1 + i)^1 + \cdots + A(1 + i)^{n-2} + A(1 + i)^{n-1}$$

This is geometric progression, the value of which is obtained by the formula:

$$S = \frac{a(r^n - 1)}{r - 1}$$

where

a is the first term = A;
r is the common ratio = $(1 + i)$; and
S is the sum of the terms = F.

hence,

$$F = A \left[\frac{(1 + i)^n - 1}{1 + i - 1} \right] = A \left[\frac{(1 + i)^n - 1}{i} \right]$$

Sometimes, knowing the derivation of a formula is helpful in the solution of a problem.

PROBLEM 9-1. COMPOUND INTEREST

What is the value at the end of 10 years of $5,000 deposited today at 6% interest?

Solution:

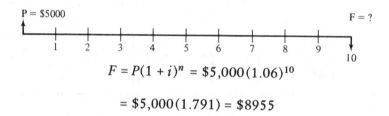

$$F = P(1 + i)^n = \$5,000(1.06)^{10}$$

(answer)
$$= \$5,000(1.791) = \$8955$$

If your answer is correct, go to the next problem.
If your answer is not correct, review pp. 234–235.

PROBLEM 9-2. PRESENT VALUE

What is the present value of a $10,000 payment to be made eight years from now, assuming interest is 10%?

Solution:

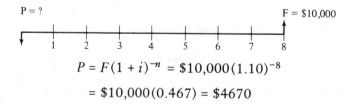

$$P = F(1 + i)^{-n} = \$10,000(1.10)^{-8}$$

(answer)
$$= \$10,000(0.467) = \$4670$$

If your answer is correct, go to the next problem.
If your answer is not correct, review pp. 234–235.

PROBLEM 9-3. EFFECTIVE INTEREST RATES

A person obtains a 36-month, $7500 home improvement loan. If the effective annual interest rate is 11%, what should the monthly payments be?

Solution:

$$i = \left(1 + \frac{j}{m}\right)^m - 1$$

where $m = 12$

$i = 0.11$

Rewriting in terms of j and solving,

$$j = m[(1 + i)^{1/m} - 1] = 12[(1.11)^{1/12} - 1] = 0.10481513 \text{ or } 10.48\% \text{ nominal rate}$$

For 36 monthly payments:

$$P = A[PVA]_{i=j/12=0.87346\%}^{n=36}$$

$$PVA = \left[\frac{(1 + i)^n - 1}{i(1 + i)^n}\right] = \left[\frac{(1.0087346)^{36} - 1}{0.0087346\,(1.0087346)^{36}}\right] = 30.775$$

$$A = \frac{7500}{30.775} = \$243.70 \qquad \text{(answer)}$$

If your answer is correct, go on to the next problem.
If your answer is not correct, review p. 233.

PROBLEM 9-4. FUTURE VALUE OF ANNUITY

What is the value at the end of 5 years of a series of annual deposits of $500 at 6% interest?

Solution:

$$F = 500\left[\frac{(1+.06)^5 - 1}{.06}\right]$$

$$F = A\left[\frac{(1 + i)^n - 1}{i}\right] = A[FVA]_{i=6\%}^{n=5}$$

$$= \$500(5.637) = \$2819 \qquad \text{(answer)}$$

If your answer is correct, go to the next problem.
If your answer is not correct, review pp. 234–235.

PROBLEM 9-5.   PRESENT VALUE OF ANNUITY

What is the uniform annual payment required to amortize a debt of $10,000 in 6 years with interest at 8%?

Solution:

P = $10,000

$$P = A\left[\frac{(1+i)^n - 1}{i(1+i)^n}\right] = A\,[PVA]_{i=8\%}^{n=6}$$

In this case, the present value is given, and the problem is to determine the annual payment required to retire the debt, with interest.

$$\$10,000 = A(4.623)$$

(answer)
$$\frac{\$10,000}{4.623} = \$2161$$

If your answer is correct, go on to Problem 9-6.
If your answer is not correct, review pp. 234–235.

PROBLEM 9-6.   FOREIGN INVESTMENT ALTERNATIVES

A corporation has the option of investing some of its excess capital in one of two possible new foreign branches, Venezuela or Greece. Estimated costs and revenues are as follows:

	Venezuela	Greece
Initial cost	125,000	150,000
Annual cost	46,000	44,000
Annual revenues	65,000	70,000

Which location would produce the highest yield on investment over a twenty-year period? Assume zero salvage value of the equipment.

Solution:

Venezuela
   Annual revenue less costs = $65,000 – $46,000 = $19,000
   Capital recovery of investment = $125,000 × CRF = $19,000

$$CRF = \frac{19,000}{125,000} = 0.152 = \frac{1}{PVA}$$

$$PVA = \frac{1}{0.152} = 6.58,\text{ which lies between 14\% and 15\% at twenty years}$$

Greece
  Annual revenue less costs = $70,000 – $44,000 = $26,000
  Capital recovery of investment = $150,000 $\times$ CRF = $26,000

$$CRF = \frac{26,000}{150,000} = 0.173 = \frac{1}{PVA}$$

$$PVA = \frac{1}{0.173} = 5.769, \text{ which lies between 16\% and 17\% at twenty years}$$

Thus, it is seen that Greece is the better choice.   (answer)

  If your solution is correct, go on to Problem 9-7.
  If your solution is not correct, review pp. 234–235.

PROBLEM 9-7. UNIFORM GRADIENT SERIES

Five annual deposits are made into a sinking fund earning 6% interest. The first year-end deposit is $10,000, and each subsequent deposit is $2,000 more than the preceding one. What will be the value of the sinking fund after the last payment?

Solution:

This problem may be solved in two different ways. One way is by evaluating the interest earned by each deposit as follows:

$F = \$10,000(1.06)^4 + \$12,000(1.06)^3 + \$14,000(1.06)^2 + \$16,000(1.06) + \$18,000$

  $= 12624.77 + 14292.19 + 15,730.40 + 16,960.00 + 18,000$

  $= \$77,607.36$ (answer)

Another way is by using the following formula:

$$F = \left(A_1 + \frac{D}{i}\right)[FVA]_{i=6\%}^{n=5} - n\frac{D}{i}$$

where

  A_1 is the first of the series of payments;
  D is the constant difference between the payments.

Thus,

$$F = \left(\$10,000 + \frac{\$2000}{0.06}\right)(5.637) - \frac{5\,(\$2000)}{0.06}$$

  $= \$244,274.03 - 166,666.67 = \$77,607.36$ (answer)

This uniform gradient series formula is based on arithmetic progression. There have been interest tables published for uniform gradient series, but their use is rather restricted.

If your answer is correct, go to Problem 9-8.
If your answer is not correct, review above solution.

PROBLEM 9-8. LAND INVESTMENT, ALTERNATIVE ASSUMPTIONS

It has been determined that 100 acres of land will be needed for future expansion ten years after construction of the present facilities. Purchase price of the land is currently $5000 per acre. Taxes are 2% per year. The land may be rented for the next ten years yielding 3% of its value. If money is worth 12% per year, what must the land be worth ten years hence to justify its purchase now?

Solution:

$$P = \$500,000$$

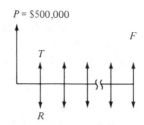

(answer)

$$F = P[CI]_{i=(12+2-3)\%}^{n=10} = \$500,000[2.839] = \$1,419,500$$

This assumes rent and taxes increase each year at the same rate as the land value. If taxes and rent remain constant, then the result would be:

$$F = P[CI]_{i=12\%}^{n=10} + (0.02 - 0.03)P[FVA]_{i=12\%}^{n=10}$$

(answer)

$$= \$500,000\,(3.106) - \$5,000\,(17.55) = \$1,465,250$$

This problem illustrates the effect variables such as rent and taxes can have on the results. Assumptions are very important.

If your solution is correct, go on to Problem 9-9.
If your solution is not correct, review pp. 234–237.

PROBLEM 9-9. ENDOWMENT FUND

A college alumnus wishes to endow a portion of a new building in perpetuity. It is estimated that the portion of costs to be supported by the endowment fund include:

$100,000 for initial construction;
$25,000 per year for operating expenses; and
$15,000 every four years for replacement and new equipment.

What amount of money must the alumnus contribute now to establish the fund? Assume an interest rate of 6%.

Solution:

$$P = \$100,000 + \$25,000\,[PVA]_{i=6\%}^{n=\infty} + \$15,000\,[PVA]_{i=6\%}^{m=\infty}$$

where

  n is number of annual payments;
  m is number of quadrennial payments.

Since interest tables do not show years to infinity, the value of PVA must be evaluated from the basic formula as follows:

$$[PVA]_{i=6\%}^{n=\infty} = \left[\frac{1 - (1+i)^{-n}}{i}\right]_{i=6\%}^{n=\infty} = \frac{1-0}{i} = \frac{1}{0.06} = 16.67$$

Converting the $15,000 quadrennial payment to an annual basis,

$$F = A\,[FVA]_{i=6\%}^{n=4}, \quad A = \frac{\$15,000}{4.375} = \$3428.57$$

Now, solving the original equation,

$$P = \$100,000 + \$25,000\,(16.67) + \$3428.57\,(16.67)$$

$$= \$100,000 + \$416,666.67 + \$57,142.86 = \$573,809.53 \qquad \text{(answer)}$$

  If your answer is correct, go to Problem 9-10.
  If your answer is not correct, review pp. 239–240.

PROBLEM 9-10. RATE OF RETURN

The initial cost of a prospective business investment consisting of plant, property, and equipment is $100,000. Annual income from the investment is expected to be $12,000, and annual expenses are expected to be $5,000. It is expected that the value of the plant, property, and equipment will be worth $75,000 after 15 years. What is the prospective rate of return (interest rate)?

Solution:

Drawing a time–payment chart:

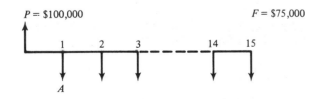

$$A = \text{Income} - \text{Expenses} = \$12,000 - \$5,000 = \$7,000$$

Relate all values to the same point in time (such as the future) and determine the interest rate that yields income plus salvage value equal to the value of the $100,000 investment if it were placed in a savings account for 15 years earning the same amount of interest compounded annually.

$$\$100,000\,[CI]_{i=?}^{n=15} = \$7,000\,[FVA]_{i=?}^{n=15} + \$75,000$$

or

$$\$75,000 = \$100,000 \, [CI]_{i=?}^{n=15} - \$7,000 \, [FVA]_{i=?}^{n=15}$$

The easiest method is trial and error, guessing at an interest rate.

Attempt 1. $i = 5.5\%$.

$$\$100,000 \, (2.232) - \$7,000 \, (22.41) = \$66,300$$

Attempt 2. $i = 6\%$

$$\$100,000 \, (2.397) - \$7,000 \, (23.28) = \$76,700$$

(answer) From the preceding calculations it is seen that the prospective rate of return is slightly less than 6%.

This is similar to $100,000 placed in a savings account at 6% interest with $7,000 withdrawn annually. The final amount remaining in the account after 15 years is nearly $75,000. As the calculations demonstrate, if the interest rate is lower, less remains at the end, and if the interest rate is higher, more remains.

If your answer is correct, go on to Problem 9-11.
If your answer is not correct, review pp. 235–236.

PROBLEM 9-11. BOND ISSUE

A county government wishes to issue $1 million worth of bonds paying 6% annual interest and maturing in 25 years. A sinking fund is to be established to pay off the bonds at maturity. The sinking fund will earn 5% interest. Determine the annual cost of the bond issue.

Solution:

The county must make two types of annual payments: (1) payment of annual interest to the bond holders, and (2) payment into a sinking fund whose value after 25 years is equal to the face value of the bonds, i.e., $1 million. The sum of these two payments is the total annual cost of the bond issue.

$$A_1 = \$1,000,000 \, (0.06) = \$60,000$$

$$F = A_2 \, [FVA]_{i=5\%}^{n=25} = A_2 (47.73), \, F = \$1,000,000$$

$$\therefore A_2 = \frac{\$1,000,000}{47.73} = \$20,951$$

(answer)

$$A = A_1 + A_2 = \$60,000 + \$20,951 = \$80,951$$

If your answer is correct, go to Problem 9-12.
If your answer is not correct, review p. 237.

PROBLEM 9-12. COLLEGE EDUCATION FUND

An engineer desires to establish a college fund for her new child's college education. She estimates the needs will be $4000 on the daughter's 18th, 19th, 20th, and 21st birthdays. The fund is to receive a lump sum of $2000 on the day of the daughter's birth and a fixed amount on the child's first through 17th birthdays, inclusive. If the fund earns 5% per annum, what should the annual deposit be?

Solution:

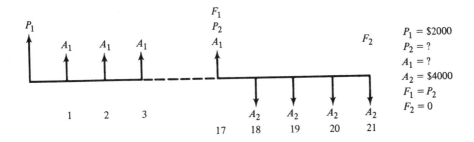

$$P_1 = \$2000$$
$$P_2 = ?$$
$$A_1 = ?$$
$$A_2 = \$4000$$
$$F_1 = P_2$$
$$F_2 = 0$$

This problem is best worked in two parts. First, calculate the amount needed (P_2) at the end of the 17th year to make the $4000 annual payments ($A_2$) for the next four years. Then calculate the size of the annual deposits (A_1) for the first 17 years which will yield the future amount (F_1) needed to make the four $4000 payments.

$$P_2 = \$4000 \, [PVA]_{i=5\%}^{n=4} = \$4000 \, (3.546) = \$14{,}184$$

$$F_1 = P_1 \, [CI]_{i=5\%}^{n=17} + A_1 \, [FVA]_{i=5\%}^{n=17} = \$14{,}184$$

$$A_1 (25.84) = \$14{,}184 - \$2000 \, (2.292) = \$9{,}600$$

$$A_1 = \frac{\$9600}{25.84} = \$371.52 \qquad \text{(answer)}$$

If your answer is correct, go to Problem 9-13.
If your answer is not correct, review pp. 235–237.

PROBLEM 9-13. RETIREMENT PLAN

A man joins a company on his fortieth birthday at a salary of $30,000. The company retirement program pays one-half the average salary over the last five years of employment, starting at age 65. Vesting occurs after ten years of employment, after which time he may leave the company. Life expectancy is 75 years, and interest earned on the pension fund is assumed to be 9%. The company pays half the cost of the retirement plan. The man's salary is expected to increase at an average rate of $4,000 per year.

Determine the man's annual contribution to his retirement program to ensure adequate funding after vesting.

Solution:

$P_1 = \$30,000$

$F_1 = P_2 = \$30,000 + \$40,000 = \$70,000$

Average salary for the last five years = \$62,000

Annual income from retirement plan = \$31,000

The employee must contribute enough to support half of \$31,000, or \$15,500, annually after his sixty-fifth birthday.

$$P_3 = \$15,500\,[PVA]_{i=9\%}^{n=10} = \$15,500\,(6.418) = \$99,479 = F_2$$

$$P_2 = F_2\,[PV]_{i=9\%}^{n=15} = \$99,479\,(0.2746) = \$27,314.39 = F_1$$

This is the amount the worker needs to contribute to the fund after ten years with the company.

Using the uniform gradient series, because the salary increase is assumed to be a uniform \$4,000 per year:

$$F_1 = \left(A_1 + \frac{D}{i}\right)[FVA]_{i=9\%}^{n=10} - n\,\frac{D}{i}$$

In terms of A_1, $D = \dfrac{4,000}{30,000}\,A_1 = 0.1333\,A_1$

$$\$27,314.39 = \left(A_1 + \frac{0.1333\,A_1}{0.09}\right)(15.19) - \frac{1.333\,A_1}{0.09}$$

$$= A_1\left(15.19 + 15.19 \times \frac{0.1333}{0.09} + \frac{1.333}{0.09}\right)$$

$$= 22.88\,A_1$$

$$A_1 = \frac{\$27,314.39}{22.88} = \$1,193.87,$$

the man's first-year contribution to his retirement plan. Since the annual contribution increases at a uniform rate in direct proportion to his income, the fraction of his annual salary to be withheld is:

(answer)

$$\frac{A_1}{P_1} = \frac{\$1,193.87}{30,000} = 0.0398$$

If your answer is correct, go on to Problem 9-14.
If your answer is not correct, review Problem 9-7.

244

PROBLEM 9-14. BOND VALUATION

A corporation desires to build and furnish a new office building at an estimated cost of $5,275,000. It is offering bonds paying 7% interest compounded semi-annually and due 10 years hence. Assuming prospective buyers wanted 9% return, what should they offer for the bonds?

Solution:

Assume the bond has a face value of $1,000.

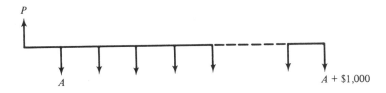

7% interest compounded semi-annually amounts to an effective annual interest rate i of:

$$i = \left(1 + \frac{0.07}{2}\right)^2 - 1 = (1.035)^2 - 1 = 0.07123 \quad \text{or} \quad 7.123\%$$

Thus, $A = \$1,000\,(0.07123) = \$71.23$ annually.

Solving for the present value of such a return in which the desired interest rate is 9%:

$$P = \$1,000\,[PV]_{i=9\%}^{n=10} + \$71.23\,[PVA]_{i=9\%}^{n=10}$$

$$= \$1,000\,(0.42241) + \$71.23\,(6.418) = \$422.41 + \$457.15 = \$879.66 \qquad \text{(answer)}$$

Thus, the bond buyers would offer $879.66 for each $1,000 bond sold by the company, and the company would need to sell 6,000 bonds to raise the needed capital.

If your answer is correct, continue to the next section.
If your answer is not correct, review pp. 235–237.

DEPRECIATION

Introduction

Depreciation is an important factor in practical economic situations. It is that portion of asset reduction attributable to a given period of time, usually one year. Two factors are required in depreciation calculations: a depreciation method and an asset life.

Depreciation methods include straight-line and accelerated methods such as sum-of-the-years-digits and declining balance. The straight-line method used to be the most popular because it was so simple. However, the accelerated methods are popular now because they offer more rapid write-offs for tax purposes, and, with the use of computers, computations are simple.

Asset life is determined by tradition, coupled with whatever the IRS will accept; for the taxpayer, the shorter the life, the better. Actually, asset life is intended to be the number of years that the asset has value.

Depreciation Methods

The three most popular depreciation methods are described below. Application of these methods is demonstrated by problems in subsequent pages.

Straight-Line.

$$\text{Annual depreciation} = \frac{P - L}{n}$$

where

P = acquisition cost;
L = estimated salvage value after n years;
n = asset life in years.

Sum-of-the-Years-Digits. If an asset has an expected life of five years, the "years digits" are 1, 2, 3, 4, and 5; and their sum is 15. The years digits are used in their reverse order as numerators and their sum as the common denominator of a series of fractions. The depreciation charge for the first year is thus arbitrarily defined as $\frac{5}{15}$ of the difference between acquisition cost and estimated salvage value (P - L), and so on to $\frac{1}{15}$ of this amount (P - L) in the fifth year.

Declining Balance. The appropriate depreciation rate r is obtained by the formula:

$$r = 1 - (L/P)^{1/n}$$

The depreciation rate is then applied to the remaining value at the end of each year to obtain a new book value. A positive salvage value must be assumed because an amount cannot be reduced to zero by successive applications of a fixed rate. Minor changes in L, however, will have greatly exaggerated effects on r, and hence, will yield large differences in the depreciation charges for the early years. Consequently, practical application of the method generally requires selection of a reasonable rate, with the result that the salvage value becomes an arbitrary residual figure. In this case, then, the formula becomes:

$$L = P(1 - r)^n$$

where r is some arbitrary, reasonable rate.

For double declining balance (DDB), the rate is twice the corresponding straight-line rate of $1/n$. Thus, an asset with a 10-year life is depreciated at 20% ($r = 2/n$) of its undepreciated balance each year. Since the asset is never 100% depreciated this way, the undepreciated portion of the account balance is written off the last year.

Comparison of Methods

The chart below compares the effects of the preceding methods. An asset costing $6,000 with a five-year life and a salvage value of approximately $800 is assumed. The declining balance rate is 1/3, as calculated by the previously given formula.

246

Periods	Straight-Line		Declining Balance ($r = \frac{1}{3}$, or $\frac{5}{3}$ DB)*		Sum-of-the-Years-Digits	
	beginning balance	period charge	beginning balance	period charge	beginning balance	period charge
Year 1	$6000	$1040	$6000	$2000	$6000	$1733
Year 2	4960	1040	4000	1333	4267	1387
Year 3	3920	1040	2667	889	2880	1040
Year 4	2880	1040	1778	593	1840	693
Year 5	1840	1040	1185	395	1147	347
Balance	800		790		800	
Total amortized		$5200		$5210		$5200

*The rate r is related to the declining balance prefix (i.e., XDB, where X is some prefix number, such as 2 for DDB) by the relation:

$$r = \frac{X}{n}$$

For example, if $r = \frac{1}{3}$ and $n = 5$, then $X = \frac{5}{3}$. This depreciation is then termed five-thirds declining balance, or $\frac{5}{3}$ DB. Or, if 5 years SL is 20% per year, than 5 years DDB is 2 × 20%, or 40% per year of remainder.

EQUIVALENT COSTS

Introduction

Now that the basic tools of engineering economics—interest and depreciation—have been defined and demonstrated, you can solve more complex problems which involve converting various nonequivalent costs to an equivalent-cost basis for comparison of alternative plans. These comparisons may be made on an annual-cost basis, a present-value basis, or a future-value basis. The basis selected is up to the discretion of the individual. One basis is usually obviously easier to compute than the others, so it is selected. Generally, alternative cost comparisons may be made on any equivalent cost basis.

PROBLEM 9-15. EQUIVALENT ANNUAL COSTS

A power company is planning to add 50,000 KW to its electrical energy production capacity. It has two alternatives available: steam (using coal) and hydroelectric. Annual costs for the two systems are estimated as follows:

	Steam	Hydroelectric
Initial investment	$4,000,000	$9,000,000
Interest on investment	7%	7%
Taxes	6%	6%
Operating costs:		
coal	$8 per ton	—
other	$150,000	$80,000
Maintenance	$100,000	$60,000
Depreciation	$125,000	$150,000

One ton of coal produces 2500 KWH of electrical energy. It is expected that the installation will operate at 60% of capacity on an annual basis.

Determine which alternative will produce the lower cost energy.

Solution:

For costs that cannot be combined directly, use the technique of equivalent annual cost to convert all costs that are not on an annual basis to annual costs. Each alternative has a first cost and annual costs.

Steam

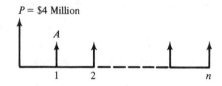

Life n is:

$$n = \frac{\$4,000,000}{\$125,000} = 32 \text{ years}$$

Fixed annual costs are:

$$A_1 = \$4,000,000 \ (0.13) + \$150,000 + \$100,000 + \$125,000 = \$895,000$$

Annual variable costs (60% use factor) is:

$$\frac{\$8}{\text{ton}} \times \frac{\text{ton}}{2500 \text{ KWH}} \times \frac{8760 \text{ hrs}}{\text{year}} \times 0.6 \times 50,000 \text{ KW} = \$840,960$$

Total equivalent annual cost is:

(answer) $\$895,000 + \$840,960 = \$1,735,960$

Hydroelectric

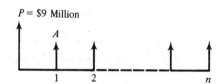

Life, n, is:

$$n = \frac{\$9,000,000}{150,000} = 60 \text{ years}$$

Fixed annual costs are:

$$A_1 = \$9,000,000 \ (0.13) + \$80,000 + \$60,000 + \$150,000 = \$1,460,000$$

Total equivalent annual cost is:

$$\$1,460,000$$ (answer)

Comparing the two alternatives, it is seen that hydroelectricity is cheaper. (answer)

If your answers are correct, go to Problem 9-16.
If your answers are not correct, review pp. 237–238.

PROBLEM 9-16. EQUIVALENT ANNUAL COST (FUTURE VALUE)

The director of parks and recreation of a city has been allocated a budget over the next ten years to convert some open space to recreational use and to maintain other existing parks and recreation areas. He expects that his expenditures will amount to $15,000 per year at the end of each of the next four years and $25,000 per year at the end of each of the following six years. He also expects to spend lump sums of $30,000 at the end of five years and $60,000 at the end of nine years.

What should his annual budget be in equivalent equal annual expenditures over the next ten years? Assume money is worth 8%.

Solution:

This problem is best solved by determining the equivalent future value of all costs and then converting to an equivalent annual cost. To simplify the calculations, assume A_1 and A_2 are equivalent to an annual cost of $15,000 for 10 years and $10,000 for the last six years, respectively.

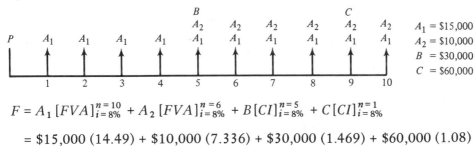

$$F = A_1 [FVA]_{i=8\%}^{n=10} + A_2 [FVA]_{i=8\%}^{n=6} + B [CI]_{i=8\%}^{n=5} + C [CI]_{i=8\%}^{n=1}$$

$$= \$15,000 (14.49) + \$10,000 (7.336) + \$30,000 (1.469) + \$60,000 (1.08)$$

$$= \$399,580$$

Converting this to equivalent annual cost:

$$F = A [FVA]_{i=8\%}^{n=10} = A(14.49) = \$399,580$$

$$A = \frac{\$399,580}{14.49} = \$27,576 \text{ per year}$$ (answer)

If your answer is correct, go to Problem 9-17.
If your answer is not correct, review pp. 235–238.

PROBLEM 9-17. SALVAGE VALUE

A city engineer requires a new pump for use by city workmen in maintaining free-flowing sewer lines. The pump will be used daily. Two alternatives are available: a used diesel-powered pump or a new gasoline-powered pump. Both pumps are expected to provide service for five years.

Determine which pump is the most economical to purchase if the interest rate is 10%.

Cost and salvage data are as follows:

	Gasoline	Diesel
Initial cost	$3,000	$5,000
Annual fuel cost	2,000	800
Annual maintenance	200	100
Salvage value	none	1,000

Solution:

In the situation where there is a salvage value involved, the formula,

$$A = \frac{(P - L)}{[PVA]_i^n} + Li$$

P - initial cost
L - lost value

may be used to convert the initial cost and salvage value to an uniform annual series.

Gasoline

P = $3,000

A A A A A

i = 10%
n = 5 Years

Equivalent annual cost of the initial cost is:

$$P = A[PVA]_{i=10\%}^{n=5} = A[3.791] = \$3,000$$

$$A = \frac{\$3,000}{3.791} = \$791$$

Total equivalent annual cost is:

(answer)

$$\$791 + \$2,000 + \$200 = \$2991$$

Diesel

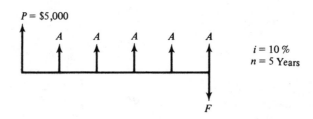

P = $5,000

A A A A A

i = 10 %
n = 5 Years

F

Equivalent annual cost of the initial cost and the salvage value is:

$$A = \frac{P - L}{[PVA]_{i=10\%}^{n=5}} + Li = \frac{5,000 - 1,000}{[3.791]} + 1,000\,(0.1)$$

$$= \$1,055 + \$100 = \$1,155$$

Total equivalent annual cost is:

$$\$1,155 + \$800 + \$100 = \$2,055 \qquad \text{(answer)}$$

Thus, it is seen that the diesel-powered pump is the better investment. (answer)

Alternative Solution:

An alternative method of converting a first cost and salvage value to a uniform annual series is an approximate method based upon straight-line depreciation and average interest. The following formula is used:

$$A = \frac{(P - L)}{n} + (P - L)\left[\frac{i}{2}\right]\left[\frac{n + 1}{n}\right] + Li$$

This method suffers from the following disadvantages:

1. it requires more calculations (you cannot use tables);
2. it is less accurate;
3. it should not be used where there is a difference in lives for the alternatives.

Gasoline

$$A = \frac{3,000}{5} + 3,000\left[\frac{0.1}{2}\right]\left[\frac{6}{5}\right] = \$780 \qquad \text{(answer)}$$

This compares with $791 for the more accurate method.

Diesel

$$A = \frac{5,000 - 1,000}{5} + (5,000 - 1,000)\left[\frac{0.1}{2}\right]\left[\frac{6}{5}\right] + 1,000\,(0.1) = \$1,140$$

This compares with $1,155 for the more accurate method.

If your answers are correct, go to Problem 9-18.
If your answers are not correct, review the above solutions.

PROBLEM 9-18. EQUIVALENT ANNUAL COST (DIFFERENT LIVES)

The Atlantic, Mohawk & Phoenix Railroad needs to upgrade its road bed. One of the items requiring replacement is a bridge over a canal. Two alternatives must be considered:

1. Build a suspension bridge high over the canal so that ships, including large ore boats, can pass underneath; this requires relocation of the road bed approach to a higher level.
2. Build a lift bridge that must raise to accommodate passing ships; this requires no track relocation and can be constructed at a lower cost but requires an attendant on duty at all times.

Using the data listed below, which of the two alternatives would have a lower annual cost? Assume an 8% interest rate.

	Suspension Bridge	Lift Bridge
Initial cost	$1,500,000	$500,000
Annual maintenance (including bridge operator)	20,000	50,000
Special maintenance every 4 years	none	25,000
Life	permanent	60 years
Salvage value	none	100,000

Solution:

This is a situation where the lives of the two alternatives are different. By comparing on an equivalent annual cost basis, the fact that the lives are different has no effect.

Suspension Bridge

$P = \$1,500,000$ $i = 8\%$ $n = \infty$

Equivalent annual cost of the initial cost is:

$$P = A\,[PVA]_{i=8\%}^{n=\infty} = \frac{A}{i} = \frac{A}{0.08} = 12.5A = \$1,500,000$$

$$A = \frac{\$1,500,000}{12.5} = \$120,000$$

Total equivalent annual cost is:

(answer) $\$120,000 + \$20,000 = \$140,000$

Lift Bridge

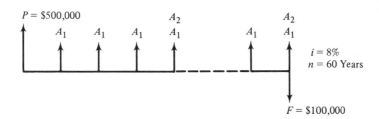

Equivalent annual cost of the initial cost is:

$$P = A \, [PVA]_{i=8\%}^{n=60} = A(12.38) = \$500,000$$

$$A = \frac{\$500,000}{12.38} = \$40,388$$

Equivalent annual income of the salvage value is:

$$F = A \, [FVA]_{i=8\%}^{n=60} = A[1253] = \$100,000$$

$$A = \frac{\$100,000}{1253} = \$80$$

Equivalent annual cost of the special extra maintenance is:

$$F = A \, [FVA]_{i=8\%}^{n=4} = A[4.506] = \$25,000$$

$$A = \frac{\$25,000}{4.506} = \$5,548$$

Total equivalent annual cost is:

$$\$40,388 - \$80 + \$5548 + \$50,000 = \$95,856 \qquad \text{(answer)}$$

Thus, it is seen that the lift bridge is cheaper. (answer)

If your answers are correct, go to Problem 9-19.
If your answers are not correct, review pp. 237 and 238.

PROBLEM 9-19. PRESENT WORTH (ALTERNATIVE INVESTMENTS)

An electric power company must expand its service. Two alternatives are available to the company, both of which will be adequate for 25 years. The table below shows the costs of each alternative over its lifetime.

	Alternative A	*Alternative B*
Initial investment	$ 75,000	$150,000
Secondary investment after 5 years	100,000	—

	Alternative A	Alternative B
Annual taxes and maintenance:		
years 1–25	–	900
years 1–5	400	–
years 6–25	1,100	–

Equate the two alternatives on a present worth basis. Assume cost of money is 9%.

Solution:

Alternative A

$$P = \$75,000 + \$100,000[PV]_{i=9\%}^{n=5} + \$1,100[PVA]_{i=9\%}^{n=25} - \$700[PVA]_{i=9\%}^{n=5}$$

$$= \$75,000 \ \$100,000 \ (0.65) + \$1,100 \ (9.8226) - \$700 \ (3.89)$$

(answer)

$$= \$75,000 + \$65,000 + \$10,805 - \$2,723 = \$148,082$$

Alternative B

$$P = \$150,000 + \$900[PVA]_{i=9\%}^{n=25} = \$150,000 + \$900 \ (9.8226)$$

(answer)

$$= \$150,000 + \$8,840 = \$158,840$$

(answer) Alternative A is the better investment.

If your answers are correct, go to Problem 9-20.
If your answers are not correct, review pp. 235–238.

PROBLEM 9-20. PRESENT WORTH (RATE OF RETURN)

A limited liability partnership has an opportunity to extend its services to a new geographical area. Projected income and expenses for the two locations are summarized below:

	Location A	Location B
Initial investment	$125,000	$150,000
Annual expenses	48,000	43,000
Annual revenues	60,000	65,000

The life of either investment is estimated at 15 years.
Determine which location would yield the highest return on investment.

Solution:

Location A

$$P = \$125,000 = (\$60,000 - \$48,000) \ [PVA]_{i=?}^{n=15}$$

$$[PVA]_{i=?}^{n=15} = \frac{125,000}{12,000} = 10.42$$

From the interest tables, $i \approx 5\%$ (answer)

Location B

$$P = \$150,000 = (\$65,000 - \$43,000) \, [PVA]_{i=?}^{n=15}$$

$$[PVA]_{i=?}^{n=15} = \frac{150,000}{22,000} = 6.82 \qquad \text{(answer)}$$

From the interest tables, $i \approx 12\%$. Thus, Location B is clearly the better choice, even (answer) though the initial investment is somewhat higher.

If your answers are correct, proceed to the next section.
If your answers are not correct, review p. 238.

BREAKEVEN ANALYSIS

The breakeven concept is used in economic analysis to choose between alternatives as well as to find minimum production levels. This technique is useful where an unknown parameter such as capacity or production level is involved.

PROBLEM 9-21. ALTERNATIVE METHODS

An electronic circuit can be implemented using discrete parts or using large scale integration (LSI). Production setup for the discrete implementations cost \$100; the cost for each circuit in \$40. The LSI implementation has an initial mask charge of \$2,000, a production setup cost of \$50, and a variable piece-part cost of \$5.
Determine the production volume required to justify going to LSI.

Solution:

Calculate the cost C of making Q parts by each method.

Discrete

$$C_d = 100 + 40Q$$

LSI

$$C_l = 2000 + 50 + 5Q$$

The quantity Q where either method has the same cost occurs where $C_d = C_l$. Therefore:

$$100 + 40Q = 2,050 + 5Q$$

$$35Q = 1,950$$

$$Q = \frac{1,950}{35} = 55.71$$

(answer)   A production volume of at least 56 circuits is required to justify using LSI.

If your answer is correct, go to Problem 9-22.
If your answer is not correct, review the above solution.

PROBLEM 9-22. PRODUCTION CAPACITY

A tree-removal firm is considering the purchase of power equipment to facilitate the removal and chipping of diseased elm trees. The cost of the equipment is $24,000. Its life is expected to be 10 years with a $3,000 salvage value. Total fixed costs are estimated to be $1,000 annually; daily variable costs are estimated to be $100. The equipment can be used to remove 15 trees per day. At the present time, trees are removed manually requiring an average of 2 days per tree per worker at a cost of $15 per day.

How many trees must be removed per year to make the purchase of the equipment economically feasible? Assume 10% to be a reasonable rate of return.

Solution:

Manual Method

Trees removed manually cost $30 each. The cost to remove trees manually for 10 years, working X days per year, is:

$$F_M = \left[\frac{\$30}{\text{tree}} \times \frac{N \text{ trees}}{\text{year}} \right] [FVA]_{i=10\%}^{n=10} = \$30N \, (15.94) = \$478.2N$$

Assuming 15 trees daily,

$$\frac{N \text{ trees}}{\text{year}} = \frac{X \text{ days}}{\text{year}} \times \frac{15 \text{ trees}}{\text{day}}, \quad N = 15X$$

$$F_M = (\$478.2) \, (15X) = \$7173X$$

Automated Method

The cost to remove trees for 10 years using the new equipment is:

$$F_A = \$24,000 \, [CI]_{i=10\%}^{n=10} - \$3,000 + (\$1,000 + \$100X) \, [FVA]_{i=10\%}^{n=10}$$

$$= \$24,000 \, (2.594) - \$3,000 + (\$1,000 + \$100X) \, (15.94)$$

$$= \$75,196 + \$1,594X$$

The breakeven point occurs when $F_M = F_A$

$$\$7,173X = \$75,196 + \$1,594X$$

$$X = \frac{75,196}{5,579} = 13.48 \text{ days per year}$$

$$N = 15X = (15)(13.48) = 202.18 \text{ trees per year} \qquad \text{(answer)}$$

If your answer is correct, continue to the next section.
If your answer is not correct, review pp. 235–238.

SPECIAL PROBLEMS

Special types of engineering problems appearing in past PE Examinations and not discussed in the preceding material include:

risk analysis and probability;
expected value;
marginal (or incremental) cost analysis;
minimum cost alternative;

At this point it may be helpful to define some terms that are likely to occur in the PE exam:

Risk—Possibility of a less desirable outcome, or variance from predicted result.

Uncertainty—Well-defined capital expenditures that are justified on poorly defined prospects of cash receipts.

Variability—Tendency to scatter data (variance, σ^2).

Expected Value—Average of possible outcomes weighted according to their probability. For example: Possible results of investing in ABC Company are:

Result	$Pr\{x\}$
$300,000 gain	0.2
50,000 loss	0.8

Expected Value:

$$E = 0.2 (\$300,000) - 0.8 (\$50,000) = \$20,000$$

The following are examples of these types of problems.

PROBLEM 9-23. RISK ANALYSIS

After purchasing a small beach home for use as an office, a private consulting engineer has learned that occasionally a severe storm will cause high tides and heavy seas, inflicting considerable damage. It is estimated that the damage would amount to $20,000. Statistics show that such a storm occurs on the average of once every 5 years. The engineer has three options available:

1. Build a seawall having a life of 20 years at a cost of $35,000.
2. Purchase insurance at an annual cost, paid in advance, of $4,000 annually.
3. Pay for the damage costs out-of-pocket as they occur.

Which option should the engineer select if the cost of money is 10%?

Solution:

Compare the three options on an annual cost basis.

Option 1

$$\$35,000 = A \,[PVA]_{i=10\%}^{n=20 \text{ years}} = A \,(8.514)$$

(answer)
$$A = \frac{\$35,000}{8.514} = \$4,110.88$$

Option 2

Assuming the insurance premiums would remain the same forever, the equivalent annual cost of the first year's premium plus subsequent annual premiums is:

(answer)
$$A = \frac{\$4,000}{[PVA]_{i=10\%}^{n=\infty}} + \$4,000 = \frac{\$4,000}{10} + \$4,000 = \$4,400$$

Option 3

(answer)
$$A = \frac{\$20,000}{5} = \$4,000$$

From a purely mathematical standpoint, Option 3 has the lowest cost. However, statistics seldom work on an individual basis, and cost estimates are never exact. Therefore, a conservative person would probably spend the extra money and build the seawall. It would also alleviate the inconvenience of having to undergo the problem of recovering from the storm damage. Option 2 does not appear to be a viable alternative.

If your answers are correct, go on to Problem 9-24.
If your answers are not correct, review the above solution.

PROBLEM 9-24. EXPECTED VALUE, FLOOD PROTECTION PLAN

A factory is located where there is possibility of flood damage from a nearby river. Any of five alternate designs of varying cost may be implemented, as tabulated below. Each design has a fifty-year life, with no salvage value. Costs are on an annual basis. Interest is at 12%.

Determine the minimum cost plan for protection against flood damage.

Design	Protection Flow Rate (CFPS)	Probability of Greater Flow (P)	Average Cost of Greater Flow (X)	Cost of Design (Y)
A	1200	0.1	$450,000	0
B	2400	0.05	500,000	150,000
C	3600	0.01	700,000	250,000
D	4800	0.005	1,000,000	400,000
E	6000	0.001	1,500,000	600,000

Solution:

Design	Expected Value of Annual Cost = Risk = $(P)(X)$ +	Annual Control Cost (Capital Recovery) = $Y[CRF]_{i=12\%}^{n=50}$ =	Total Cost
A	0.1(450,000) +	0	= $45,000
B	0.05(500,000) +	150,000(0.12042) =	43,063
C	0.01(700,000) +	250,000(0.12042) =	37,105
D	0.005(1,000,000) +	400,000(0.12042) =	53,168
E	0.001(1,500,000) +	600,000(0.12042) =	73,752

Design C is the lowest cost.

If your answer is correct, go on to Problem 9-25.
If your answer is not correct, review p. 257.

PROBLEM 9-25. EXPECTED VALUE, HIGHWAY BRIDGE

The highway department needs a new bridge and can build a two-lane bridge for $200,000 or a four-lane bridge for $350,000. If a two-lane bridge is built now and later it is determined that two more lanes are needed, the additional cost will be $200,000 plus $20,000 for each year's delay. Money is worth 10%.
The probability of requiring two more lanes in the future is:

Year	Probability
3	0.1
4	0.2
5	0.3
6	0.4

Should a two-lane or a four-lane bridge be built now?

Solution:

Plan A. Build a two-lane bridge now and add more lanes if and when needed.
Plan B. Build four lanes now.
Use present-value analysis on these two plans.

Plan A

Using the concept of expected costs and the formula:

$$CE = (\text{cost}) (\text{probability of incurring cost})$$

the following table can be developed:

Year	Present-Value $(CE)\,[PV]_i^n$	Cost
0	cost for 1st two lanes = $200,000	
3	(260,000) (0.1) (0.7513) =	19,534
4	(280,000) (0.2) (0.683) =	38,248
5	(300,000) (0.3) (0.6209) =	55,881
6	(320,000) (0.4) (0.5645) =	72,256
	Total cost = $385,919	

Plan B

$$P = \$350,000$$

From the above analysis, it is seen that Plan B should be chosen.

If your answer is correct, proceed to Problem 9-26.
If your answer is not correct, review the above solution.

PROBLEM 9-26. MARGINAL COST

The table below shows the fixed and variable costs incurred in producing from 0 to 10 units of a product. Determine the maximum number of units that may be produced such that marginal cost does not exceed marginal revenue. Each unit may be sold for $1,000.

No. units	Fixed costs	Direct production costs	Indirect variable costs
0	1000	0	0
1		700	130
2		1300	200
3		1900	275
4		2500	375
5		3000	550
6		3500	750
7		4000	1100
8		4400	1500
9		4800	2150
10		5200	3000

Solution:

Marginal cost MC is the additional cost incurred in producing each additional unit. Marginal revenue MR is the additional income received from the sale of each additional unit. In this case, $MR = \$1,000$ for each sale.

260

To determine the point at which MC begins to exceed MR, complete the table as follows:

No. units	(1) Fixed costs	(2) Direct production costs	(3) Indirect variable costs	(4) Total costs (1) + (2) + (3)	Marginal cost (MC)
0	1000	0	0	1000	0
1		700	130	1830	830
2		1300	250	2550	720
3		1900	325	3225	675
4		2500	400	3900	675
5		3000	600	4600	700
6		3500	820	5320	720
7		4000	1100	6100	780
8		4400	1500	6900	800
9		4800	2100	7900	1000
10		5200	3000	9200	1300

From the completed table, it is seen that $MC = MR = \$1,000$ at the ninth unit. Therefore, any additional units are not profitable to produce since it costs more to produce them than the revenue received from their sale.

If your answer is correct, proceed to Problem 9-27.
If your answer is not correct, review the above solution.

PROBLEM 9-27. INCREMENTAL COSTS, LAMBDA DISPATCH

The incremental operating costs per KWH for three generators are given by the equations:

$$\lambda_1 = 0.3P_1 + 70$$

$$\lambda_2 = 0.35P_2 + 55$$

$$\lambda_3 = 0.45P_3 + 35$$

where P_1, P_2, and P_3 are in KWH.
The power output range in KW for each generator is:

GEN	MIN	MAX
1	750	3000
2	1000	4500
3	1250	6000

Determine the optimum load allocation among the three generators for total loads of 5000, 7500, and 10,000 KW.

Solution:

This problem requires the use of Lagrange multiplier mathematics (See *Mark's Handbook*, 7th Edition, page 17–63, bibliography reference 37). This is a lambda dispatch problem in which several generators are interconnected to provide customers with an adequate supply of electric energy. For all loads, $\lambda_1 = \lambda_2 = \lambda_3$ for optimum operation.

Solving for P_1 and P_3 in terms of P_2:

$$0.3P_1 + 70 = 0.35P_2 + 55$$

$$P_1 = \frac{0.35P_2 - 15}{0.3} = 1.167P_2 - 50$$

$$0.45P_3 + 35 = 0.35P_2 + 55$$

$$P_3 = \frac{0.35P_2 + 20}{0.45} = 0.777P_2 + 44.44$$

For 5000 KW output:

$$P_1 + P_2 + P_3 = 5000 = 1.167P_2 - 50 + P_2 + 0.777P_2 + 44.44$$

(answer) $$P_2 = \frac{5005.55}{2.944} = 1700 \text{ KW}$$

(answer) $$P_1 = 1.167(1700) - 50 = 1933.33 \text{ KW}$$

(answer) $$P_3 = 0.777(1700) + 44.4 = 1366.67 \text{ KW}$$

All loads are within limits.

For 7500 KW output:

$$P_1 + P_2 + P_3 = 7500$$

(answer) $$P_2 = \frac{7505.55}{2.944} = 2549 \text{ KW}$$

(answer) $$P_1 = 1.167(2549) - 50 = 2924 \text{ KW}$$

(answer) $$P_3 = 0.777(2549) + 44.44 = 2027 \text{ KW}$$

All loads are within limits.

For 10,000 KW output:

$$P_1 + P_2 + P_3 = 10,000$$

$$P_2 = \frac{10,005.55}{2.944} = 3398.1 \text{ KW}$$

$$P_1 = 1.167(3398.1) - 50 = 3914.5 \text{ KW}$$

This exceeds the maximum limit for generator 1. Therefore, let:

(answer) $$P_1 = 3000 \text{ KW}$$

Now, let $\lambda_2 = \lambda_3$

$$10,000 - 3000 = 7000 = P_2 + 0.777P_2 + 44.44$$

262

$$P_2 = \frac{6955.55}{1.7777} = 3912.5 \text{ KW}$$ (answer)

$$P_3 = 0.777 \, (3912.5) + 44.44 = 3087.5 \text{ KW}$$ (answer)

If your answers are correct, proceed to Problem 9-28.
If your answers are not correct, review the above solution.

PROBLEM 9-28. MINIMUM COST ALTERNATIVES

The park department of a city is considering the purchase of one of two types of lawn fertilizer, dry and liquid.

The dry type costs $25 per 100 lb and has a coverage of 30,000 square feet per 100 lb. It requires only two applications per growing season and can be applied by a workman at the rate of 20,000 square feet per hour.

The liquid type costs $15 per gallon and has a coverage of 30,000 square feet per gallon. It requires three applications per growing season and can be applied by a workman at the rate of 15,000 square feet per hour.

If the workmen are paid at the rate of $3.50 per hour, which type of fertilizer gives the lowest cost coverage?

Solution:

Dry Type

$$\frac{\$25}{100 \text{ lb}} \times \frac{100 \text{ lb}}{30,000 \text{ ft}^2} + \frac{\$3.50}{\text{hr}} \times \frac{\text{hr}}{20,000 \text{ ft}^2}$$

$$= \$8.33 \times 10^{-4} + \$1.75 \times 10^{-4} = \$1.01 \times 10^{-3} \text{ per ft}^2$$

Two applications per season cost 0.202¢/ft². (answer)

Liquid Type

$$\frac{\$15}{\text{gal}} \times \frac{\text{gal}}{30,000 \text{ ft}^2} + \frac{\$3.50}{\text{hr}} \times \frac{\text{hr}}{15,000 \text{ ft}^2}$$

$$= \$5 \times 10^{-4} + \$2.33 \times 10^{-4} = \$7.33 \times 10^{-4} \text{ per ft}^2$$

Three applications per season cost 0.22¢/ft². (answer)

The dry type is slightly less expensive. (answer)

If your answers are not correct, review the above solution.

YOU SHOULD BE READY

Now that you have completed this programmed review, you should be well prepared to take the PE Exam. The following suggestions may help you:

- Read all instructions carefully.
- Read all problems through, checking those you feel most comfortable with.
- Work the problems which are easiest for you first.
- Include sketches or diagrams if applicable; show your method of solving the problem and your computations; demonstrate that you understand the fundamental engineering principles used in your solution.
- Write clearly and neatly.
- Check your work.
- TRY TO RELAX. No candidate can answer all the questions on the PE Exam. You have reviewed thoroughly; you have adequate reference materials with you. Proceed through the exam in a systematic manner, budgeting your time. Demonstrate your expertise.

GOOD LUCK!

Appendix Tables

TABLE A-1.   Compound Interest

Compound interest $(1 + i)^n$ to determine future worth of a single amount. For present value, use reciprocal.

n	.5%	1.0%	1.5%	2.0%	2.5%	3.0%	3.5%	4.0%	4.5%	5.0%	5.5%	6.0%
1	1.005	1.010	1.015	1.020	1.025	1.030	1.035	1.040	1.045	1.050	1.055	1.060
2	1.010	1.020	1.030	1.040	1.051	1.061	1.071	1.082	1.092	1.102	1.113	1.124
3	1.015	1.030	1.046	1.061	1.077	1.093	1.169	1.125	1.141	1.158	1.174	1.191
4	1.020	1.041	1.061	1.082	1.101	1.126	1.148	1.170	1.193	1.216	1.239	1.262
5	1.025	1.051	1.077	1.104	1.131	1.150	1.168	1.217	1.246	1.276	1.307	1.338
6	1.030	1.062	1.093	1.126	1.169	1.194	1.229	1.265	1.302	1.340	1.379	1.419
7	1.036	1.072	1.110	1.149	1.189	1.230	1.272	1.316	1.361	1.407	1.455	1.504
8	1.041	1.083	1.126	1.172	1.218	1.267	1.317	1.369	1.422	1.477	1.535	1.594
9	1.046	1.094	1.143	1.195	1.249	1.305	1.363	1.423	1.486	1.551	1.619	1.689
10	1.051	1.105	1.161	1.218	1.280	1.344	1.411	1.480	1.553	1.629	1.708	1.791
11	1.056	1.116	1.178	1.243	1.312	1.384	1.460	1.539	1.623	1.710	1.802	1.898
12	1.062	1.127	1.196	1.268	1.345	1.426	1.511	1.601	1.696	1.796	1.901	2.012
13	1.067	1.138	1.214	1.294	1.379	1.469	1.564	1.665	1.772	1.886	2.006	2.133
14	1.072	1.149	1.232	1.319	1.413	1.513	1.619	1.732	1.852	1.980	2.116	2.261
15	1.078	1.161	1.250	1.346	1.448	1.558	1.675	1.801	1.935	2.079	2.232	2.397
16	1.083	1.173	1.269	1.373	1.485	1.605	1.734	1.873	2.022	2.183	2.355	2.540
17	1.088	1.184	1.288	1.400	1.522	1.653	1.795	1.948	2.113	2.292	2.485	2.693
18	1.094	1.196	1.307	1.428	1.560	1.702	1.857	2.026	2.208	2.407	2.621	2.854
19	1.099	1.208	1.327	1.457	1.569	1.754	1.923	2.107	2.308	2.527	2.766	3.026
20	1.105	1.220	1.347	1.486	1.639	1.806	1.990	2.191	2.412	2.653	2.918	3.207
21	1.110	1.232	1.367	1.516	1.680	1.860	2.053	2.279	2.520	2.786	3.078	3.400
22	1.116	1.245	1.388	1.546	1.722	1.916	2.132	2.370	2.634	2.925	3.248	3.604
23	1.122	1.257	1.408	1.577	1.765	1.974	2.206	2.465	2.752	3.072	3.426	3.820
24	1.127	1.270	1.430	1.608	1.809	2.033	2.283	2.563	2.876	3.225	3.615	4.049
25	1.133	1.282	1.451	1.641	1.854	2.094	2.363	2.666	3.005	3.386	3.813	4.292
26	1.138	1.295	1.473	1.673	1.900	2.157	2.446	2.772	3.141	3.556	4.023	4.549
27	1.144	1.308	1.495	1.707	1.948	2.221	2.532	2.883	3.282	3.733	4.244	4.822
28	1.150	1.321	1.517	1.741	1.996	2.288	2.620	2.999	3.430	3.920	4.478	5.112
29	1.156	1.335	1.540	1.776	2.046	2.357	2.712	3.119	3.584	4.116	4.784	5.418
30	1.161	1.348	1.563	1.811	2.098	2.427	2.807	3.243	3.745	4.322	4.984	5.743
31	1.167	1.361	1.587	1.848	2.150	2.500	2.905	3.373	3.914	4.538	5.258	6.088
32	1.173	1.375	1.610	1.885	2.204	2.575	3.007	3.508	4.030	4.765	5.547	6.453
33	1.179	1.389	1.634	1.922	2.259	2.652	3.112	3.648	4.274	5.003	5.852	6.841
34	1.185	1.403	1.659	1.961	2.315	2.732	3.221	3.794	4.466	5.253	6.174	7.251
35	1.191	1.417	1.684	2.000	2.373	2.814	3.334	3.946	4.667	5.516	6.514	7.686
36	1.197	1.431	1.709	2.040	2.433	2.898	3.450	4.104	4.877	5.792	6.872	8.147
37	1.203	1.445	1.735	2.081	2.493	2.985	3.571	4.268	5.097	6.081	7.250	8.636
38	1.209	1.460	1.761	2.122	2.556	3.075	3.696	4.439	5.326	6.385	7.649	9.154
39	1.215	1.474	1.787	2.165	2.620	3.107	3.825	4.616	5.566	6.705	8.069	9.704
40	1.221	1.489	1.814	2.208	2.685	3.262	3.959	4.801	5.816	7.040	8.513	10.29
42	1.233	1.519	1.869	2.297	2.821	3.401	4.241	5.193	6.352	7.762	9.476	11.56
44	1.245	1.549	1.925	2.390	2.964	3.671	4.543	5.617	6.936	8.557	10.55	12.99
46	1.258	1.580	1.984	2.487	3.114	3.895	4.867	6.075	7.574	9.434	11.74	14.59
48	1.270	1.612	2.043	2.587	3.271	4.132	5.214	6.571	8.271	10.40	13.07	16.39
50	1.283	1.645	2.105	2.692	3.437	4.384	5.585	7.107	9.033	11.47	14.54	18.42

TABLE A-1.   Compound Interest (*Continued*)

n	.5%	1.0%	1.5%	2.0%	2.5%	3.0%	3.5%	4.0%	4.5%	5.0%	5.5%	6.0%
52	1.296	1.678	2.159	2.800	3.611	4.651	5.983	7.687	9.864	12.64	16.19	20.70
54	1.309	1.711	2.234	2.913	3.794	4.934	6.409	8.314	10.77	13.94	18.01	23.26
56	1.322	1.746	2.302	3.031	3.986	5.235	6.883	8.992	11.76	15.37	20.05	26.13
58	1.335	1.781	2.372	3.154	4.183	5.553	7.354	9.726	12.85	16.94	22.32	29.36
60	1.340	1.817	2.443	3.261	4.400	5.833	7.878	10.52	14.03	18.68	24.84	32.99
65	1.383	1.909	2.632	3.683	4.978	6.830	9.357	12.80	17.48	23.84	32.46	44.14
70	1.418	2.007	2.835	4.000	5.632	7.918	11.11	15.57	21.78	30.43	42.43	59.08
75	1.454	2.109	3.055	4.416	6.372	9.179	13.20	18.95	27.15	38.83	55.45	79.00
80	1.490	2.217	3.291	4.875	7.210	10.64	15.63	23.05	31.83	49.56	72.48	105.8
85	1.528	2.330	3.545	5.383	8.157	12.34	18.62	28.04	42.16	63.25	94.72	141.6
90	1.587	2.448	3.819	5.943	9.229	14.50	22.11	34.12	52.54	80.73	123.8	189.5
95	1.606	2.574	4.114	6.562	10.44	16.58	26.26	41.51	65.47	103.0	161.9	253.5
100	1.647	2.705	4.132	7.245	11.81	18.32	31.19	50.50	81.59	131.5	211.5	339.3

Note: Tables A-1, A-2, and A-3 are from *Project Economy* by Edwin S. Roscoe, Homewood, Illinois; Richard D. Irwin, Inc., 1961. Reprinted with permission.

n	7%	8%	9%	10%	11%	12%	13%	14%	15%	16%	17%	18%
1	1.070	1.080	1.090	1.100	1.110	1.120	1.130	1.140	1.150	1.160	1.170	1.180
2	1.145	1.166	1.188	1.210	1.232	1.254	1.277	1.300	1.322	1.346	1.369	1.392
3	1.225	1.260	1.295	1.331	1.368	1.405	1.443	1.482	1.521	1.561	1.602	1.643
4	1.311	1.360	1.412	1.464	1.518	1.574	1.630	1.689	1.749	1.811	1.874	1.939
5	1.403	1.469	1.539	1.611	1.685	1.762	1.842	1.925	2.011	2.100	2.192	2.288
6	1.501	1.587	1.677	1.772	1.870	1.974	2.082	2.195	2.313	2.436	2.565	2.700
7	1.606	1.714	1.828	1.949	2.076	2.211	2.353	2.502	2.660	2.826	3.001	3.185
8	1.718	1.851	1.993	2.144	2.305	2.476	2.658	2.853	3.059	3.278	3.511	3.759
9	1.838	1.999	2.172	2.358	2.558	2.773	3.004	3.252	3.518	3.803	4.108	4.435
10	1.967	2.159	2.367	2.594	2.839	3.106	3.395	3.707	4.046	4.411	4.807	5.234
11	2.105	2.332	2.580	2.853	3.152	3.479	3.836	4.226	4.652	5.117	5.624	6.176
12	2.252	2.518	2.813	3.138	3.498	3.896	4.335	4.818	5.350	5.936	6.580	7.288
13	2.410	2.720	3.066	3.452	3.883	4.363	4.898	5.492	6.153	6.886	7.699	8.599
14	2.579	2.937	3.342	3.797	4.310	4.887	5.535	6.261	7.076	7.988	9.007	10.15
15	2.759	3.172	3.642	4.177	4.785	5.474	6.254	7.138	8.137	9.266	10.54	11.97
16	2.952	3.426	3.970	4.595	5.311	6.130	7.067	8.137	9.358	10.75	12.33	14.13
17	3.159	3.700	4.328	5.054	5.895	6.866	7.986	9.276	10.76	12.47	14.43	16.67
18	3.380	3.996	4.717	5.560	6.544	7.690	9.024	10.58	12.38	14.46	16.88	19.67
19	3.617	4.316	5.142	6.116	7.263	8.613	10.20	12.06	14.23	16.78	19.75	23.21
20	3.870	4.661	5.604	6.727	8.062	9.646	11.52	13.74	16.37	19.46	23.11	27.39

TABLE A-1. Compound Interest (*Continued*)

n	7%	8%	9%	10%	11%	12%	13%	14%	15%	16%	17%	18%
22	4.430	5.437	6.659	8.140	9.934	12.10	14.71	17.86	21.64	26.19	31.63	38.14
24	5.072	6.341	7.911	9.850	12.24	15.18	18.79	23.21	28.63	35.24	43.30	53.11
26	5.807	7.396	9.399	11.92	15.08	19.04	23.99	30.17	37.86	47.41	59.27	73.95
28	6.649	8.627	11.17	14.42	18.58	23.88	30.63	39.20	50.07	63.80	81.13	103.0
30	7.612	10.06	13.27	17.45	22.89	29.96	39.12	50.95	66.21	85.85	111.1	143.4
32	8.715	11.74	15.76	21.11	28.21	37.58	49.95	66.21	87.57	115.5	152.0	199.6
34	9.978	13.69	18.73	25.55	34.75	47.14	63.78	86.05	115.8	155.4	208.1	278.0
36	11.42	15.97	22.25	30.91	42.82	59.14	81.44	111.8	153.2	209.2	284.9	387.0
38	13.08	18.63	26.44	37.40	52.76	74.18	104.0	145.3	202.5	281.5	390.0	538.9
40	14.97	21.72	31.41	45.26	65.00	93.05	132.8	188.9	267.9	378.7	533.9	750.4
45	21.00	31.92	48.33	72.89	109.5	164.0	244.6	363.7	538.8	795.4	1170.	1717.
50	29.46	46.90	74.36	117.4	184.6	289.0	450.7	700.2	1084.	1671.	2566.	3927.
55	41.31	68.91	114.4	189.1	311.0	509.3	830.5	1348.	2180.	3509.	5626.	8985.
60	57.95	101.3	176.0	304.5	524.1	897.6	1530.	2596.	4384.	7370.		

n	20%	22%	24%	26%	28%	30%	32%	34%	36%	38%	40%	42%
1	1.200	1.220	1.240	1.260	1.280	1.300	1.320	1.340	1.360	1.380	1.400	1.420
2	1.440	1.488	1.538	1.588	1.638	1.690	1.742	1.796	1.850	1.904	1.950	2.016
3	1.728	1.816	1.907	2.000	2.097	2.197	2.300	2.406	2.515	2.628	2.744	2.863
4	2.074	2.215	2.364	2.520	2.684	2.856	3.030	3.224	3.421	3.627	3.842	4.066
5	2.488	2.703	2.932	3.176	3.436	3.713	4.007	4.320	4.653	5.005	5.378	5.774
6	2.986	3.297	3.635	4.002	4.398	4.827	5.290	5.789	6.328	6.907	7.530	8.198
7	3.583	4.023	4.508	5.042	5.629	6.275	6.983	7.758	8.605	9.531	10.54	11.64
8	4.300	4.908	5.590	6.353	7.206	8.157	9.217	10.40	11.70	13.15	14.76	16.53
9	5.160	5.987	6.931	8.005	9.223	10.60	12.17	13.93	15.93	18.15	20.66	23.47
10	6.192	7.305	8.594	10.09	11.81	13.79	16.06	18.67	21.65	25.05	28.93	33.33
11	7.430	8.912	10.66	12.71	15.11	17.92	21.20	25.01	29.44	34.57	40.50	47.33
12	8.916	10.87	13.21	16.01	19.34	23.30	27.98	35.52	40.04	47.70	56.69	67.21
13	10.70	13.26	16.39	20.18	24.76	30.29	36.94	44.91	54.45	65.83	79.37	95.44
14	12.84	16.18	20.32	25.42	31.69	39.37	48.76	60.18	74.05	90.85	111.1	135.5
15	15.41	19.74	25.20	32.03	40.56	51.19	64.36	80.64	100.7	125.4	155.6	192.5
16	18.49	24.69	31.24	40.36	51.92	66.54	84.95	108.1	137.0	173.0	217.8	273.3
18	26.62	35.85	48.04	64.07	85.07	112.5	148.0	194.0	253.3	329.5	426.9	551.0
20	38.34	53.36	73.86	101.7	139.4	190.0	257.9	348.4	468.6	627.5	836.7	1111.
22	55.21	79.42	113.6	161.5	228.4	321.2	449.4	625.6	866.7	1195.	1640.	2240.
24	79.50	118.2	174.6	256.4	374.1	542.8	783.0	1123.	1603.	2276.	3214.	4518.
26	114.5	175.9	268.5	407.0	613.0	917.3	1364.	2017.	2965.	4334.	6300.	9110.
28	164.8	261.9	412.9	646.2	1004.	1550.	2377.	3622.	5484.	8253.		
30	237.4	389.8	634.8	1026.	1646.	2620.	4142.	6503.				

Note: Tables A-1, A-2, and A-3 are from *Project Economy* by Edwin S. Roscoe, Homewood, Illinois; Richard D. Irwin, Inc., 1961. Reprinted with permission.

TABLE A-2. Future Value of Annuity

Future value of annuity $[(1 + i)^n - 1]/i$ to determine future worth of a uniform series of periodic amounts. To determine sinking fund factor, use reciprocal.

n	.5%	1.0%	1.5%	2.0%	2.5%	3.0%	3.5%	4.0%	4.5%	5.0%	5.5%	6.0%
1	1.000	1.000	1.000	1.000	1.000	1.000	1.000	1.000	1.000	1.000	1.000	1.000
2	2.005	2.010	2.015	2.020	2.025	2.030	2.035	2.040	2.045	2.050	2.055	2.060
3	3.015	3.030	3.045	3.060	3.076	3.091	3.106	3.122	3.137	3.152	3.168	3.184
4	4.030	4.060	4.091	4.122	4.153	4.184	4.215	4.246	4.278	4.310	4.342	4.375
5	5.050	5.101	5.152	5.204	5.256	5.309	5.362	5.416	5.471	5.528	5.581	5.637
6	6.075	6.152	6.230	6.308	6.388	6.468	6.550	6.633	6.717	6.802	6.888	6.975
7	7.106	7.214	7.323	7.434	7.547	7.662	7.779	7.898	8.019	8.142	8.267	8.394
8	8.141	8.286	8.433	8.583	8.736	8.892	9.052	9.214	9.380	9.549	9.722	9.897
9	9.182	9.369	9.559	9.755	9.955	10.16	10.37	10.58	10.80	11.03	11.26	11.49
10	10.23	10.46	10.70	10.95	11.20	11.46	11.73	12.01	12.29	12.58	12.88	13.18
11	11.28	11.57	11.86	12.17	12.48	12.81	13.14	13.49	13.84	14.21	14.58	14.97
12	12.34	12.68	13.04	13.41	13.80	14.19	14.60	15.03	15.46	15.92	16.39	16.87
13	13.40	13.81	14.24	14.68	15.14	15.62	16.11	16.63	17.16	17.71	18.29	18.88
14	14.46	14.95	15.45	15.97	16.52	17.09	17.68	18.29	18.93	19.60	20.29	21.02
15	15.54	16.10	16.68	17.29	17.93	18.60	19.30	20.02	20.78	21.58	22.41	23.28
16	16.61	17.26	17.93	18.64	19.38	20.16	20.97	21.82	22.72	23.66	24.64	25.67
17	17.70	18.43	19.20	20.01	20.86	21.76	22.71	23.70	24.74	25.84	27.00	28.21
18	18.79	19.61	20.49	21.41	22.39	23.41	24.50	25.65	26.86	28.13	29.48	30.91
19	19.88	20.81	21.80	22.84	23.95	25.12	26.36	27.67	29.06	30.54	32.10	33.76
20	20.98	22.02	23.12	24.30	25.54	26.87	28.28	29.78	31.37	33.07	34.87	36.79
21	22.08	23.24	24.47	25.78	27.18	28.68	30.27	31.97	33.78	35.72	37.79	39.99
22	23.19	24.47	25.84	27.30	28.86	30.54	32.33	34.25	36.30	38.51	40.86	43.39
23	24.31	25.72	27.23	28.84	30.58	32.45	34.46	36.62	38.94	41.43	44.11	47.00
24	25.43	26.97	28.63	30.42	32.35	34.43	36.67	39.08	41.69	44.50	47.54	50.82
25	26.56	28.24	30.06	32.03	34.16	36.46	38.95	41.65	44.57	47.73	51.15	54.86
26	27.69	29.53	31.51	33.67	36.01	38.55	41.31	44.31	47.57	51.11	54.97	59.16
27	28.83	30.82	32.99	35.34	37.91	40.71	43.76	47.06	50.71	54.67	58.99	63.71
28	29.97	32.13	34.48	37.05	39.86	42.93	46.29	49.97	53.99	58.40	63.23	68.53
29	31.12	33.45	36.00	38.79	41.86	45.22	48.91	52.97	57.42	62.32	67.71	73.64
30	32.28	34.78	37.54	40.57	43.90	47.58	51.62	56.08	61.01	66.44	72.44	79.06
31	33.44	36.13	39.10	42.38	46.00	50.00	54.43	59.33	64.75	70.76	77.42	84.80
32	34.61	37.49	40.69	44.23	48.15	52.50	57.33	62.70	68.67	75.30	82.68	90.89
33	35.78	38.87	42.30	46.11	50.35	55.08	60.34	66.21	72.76	80.06	88.22	97.34
34	36.96	40.26	43.93	48.03	52.61	57.73	63.45	69.86	77.03	85.07	94.08	104.2
35	38.15	41.66	45.59	49.99	54.93	60.46	66.67	73.65	81.50	90.32	100.3	111.4
36	39.34	43.08	47.28	51.99	57.30	63.28	70.01	77.60	86.16	95.84	106.8	119.1
37	40.53	44.51	48.99	54.03	59.73	66.17	73.46	81.70	91.04	101.6	113.6	127.3
38	41.74	45.95	50.72	56.11	62.23	69.16	77.03	85.97	96.14	107.7	120.9	136.9
39	42.94	47.41	52.48	58.24	64.78	72.23	80.72	90.41	101.5	114.1	128.5	145.1
40	44.16	48.89	54.27	60.40	67.40	75.40	84.55	95.03	107.0	120.8	136.6	154.8
42	46.61	51.88	57.92	64.86	72.84	82.02	92.61	104.8	118.9	135.2	154.1	176.0
44	49.08	54.93	61.69	69.50	78.55	89.05	101.2	115.4	131.9	151.1	173.6	199.8
46	51.58	58.05	65.57	74.33	84.55	96.50	110.5	126.9	146.1	168.7	195.2	226.5

TABLE A-2. Future Value of Annuity (*Continued*)

n	.5%	1.0%	1.5%	2.0%	2.5%	3.0%	3.5%	4.0%	4.5%	5.0%	5.5%	6.0%
48	54.10	61.22	69.57	79.35	90.86	104.4	120.4	139.3	161.6	188.0	219.4	256.6
50	56.65	64.46	73.68	84.58	97.48	112.8	131.0	152.7	178.5	209.3	246.2	290.3
52	59.22	67.77	77.92	90.02	104.4	121.7	142.4	167.2	197.0	232.9	276.1	328.3
54	61.82	71.14	82.30	95.67	111.8	131.1	154.5	182.8	217.1	258.8	309.4	370.9
56	64.44	74.58	86.80	101.6	119.4	141.2	167.6	199.8	239.2	287.3	346.4	418.8
58	67.09	78.09	91.44	107.7	127.5	151.8	181.6	218.1	263.2	318.9	387.6	472.6
60	69.77	81.67	92.21	114.1	136.0	163.1	196.5	238.0	289.5	353.6	433.4	533.1
65	76.58	90.94	108.8	131.1	159.1	194.3	238.8	295.0	366.2	456.8	572.1	719.1
70	83.57	100.7	122.4	150.0	185.3	230.6	288.9	364.3	461.9	588.5	753.3	967.9
75	90.73	110.9	137.0	170.8	214.9	272.6	348.5	448.6	581.0	756.7	990.1	1301.
80	98.07	121.7	152.7	193.8	248.4	321.4	419.3	551.2	729.6	971.2	1300.	1747.
85	105.6	133.0	169.7	219.1	286.3	377.9	503.4	676.1	914.6	1245.	1704.	2343.
90	113.3	144.9	187.9	247.2	329.2	443.3	603.2	828.0	1145.	1595.	2233.	3141.
95	121.2	157.4	207.6	278.1	377.7	519.3	721.8	1013.	1433.	2041.	2924.	4209.
100	129.3	170.5	228.8	312.2	432.5	607.3	862.2	1238.	1791.	2610.	3827.	5638.

Note: Tables A-1, A-2, and A-3 are from *Project Economy* by Edwin S. Roscoe, Homewood, Illinois; Richard D. Irwin, Inc., 1961. Reprinted with permission.

n	7%	8%	9%	10%	11%	12%	13%	14%	15%	16%	17%	18%
1	1.000	1.000	1.000	1.000	1.000	1.000	1.000	1.000	1.000	1.000	1.000	1.000
2	2.070	2.080	2.090	2.100	2.110	2.120	2.130	2.140	2.150	2.160	2.170	2.180
3	3.215	3.246	3.278	3.310	3.342	3.374	3.407	3.440	3.472	3.506	3.539	3.572
4	4.440	4.506	4.573	4.641	4.710	4.779	4.850	4.921	4.993	5.066	5.141	5.215
5	5.751	5.867	5.985	6.105	6.228	6.353	6.480	6.610	6.742	6.877	7.014	7.154
6	7.153	7.336	7.523	7.716	7.913	8.115	8.323	8.536	8.754	8.977	9.207	9.442
7	8.654	8.923	9.200	9.487	9.783	10.09	10.40	10.73	11.07	11.41	11.77	12.14
8	10.26	10.64	11.03	11.44	11.86	12.30	12.76	13.23	13.73	14.24	14.77	15.33
9	11.98	12.49	13.02	13.58	14.16	14.78	15.42	16.09	16.79	17.52	18.28	19.09
10	13.82	14.49	15.19	15.94	16.72	17.55	18.42	19.34	20.30	21.32	22.39	23.52
11	15.78	16.65	17.56	18.53	19.56	20.65	21.81	23.04	24.35	25.73	27.20	28.76
12	17.89	18.98	20.14	21.38	22.71	24.13	25.65	27.27	29.00	30.85	32.82	34.93
13	20.14	21.50	22.95	24.52	26.21	28.03	29.98	32.09	34.35	36.79	39.40	42.22
14	22.55	24.21	26.02	27.97	30.09	32.39	34.88	37.58	40.50	43.67	47.10	50.82
15	25.13	27.15	29.36	31.77	34.41	37.28	40.42	43.84	47.58	51.66	56.11	60.97
16	27.89	30.32	33.00	35.95	39.19	42.75	46.67	50.98	55.72	60.93	66.65	72.94
17	30.84	33.75	36.97	40.54	44.50	48.88	53.74	59.12	65.08	71.67	78.98	87.07
18	34.00	37.45	41.30	45.60	50.40	55.75	61.73	68.39	75.84	84.14	93.41	103.7
19	37.38	41.45	46.02	51.16	56.94	63.44	70.75	78.97	88.21	98.60	110.3	123.4
20	41.00	45.76	51.16	57.27	64.20	72.05	80.95	91.02	102.4	115.4	130.0	146.6
22	49.01	55.46	62.87	71.40	81.21	92.50	105.5	120.4	137.6	157.4	180.2	206.3
24	58.18	66.76	76.79	88.50	102.2	118.2	136.8	158.7	184.2	214.0	248.8	289.5
26	68.68	79.95	93.32	109.2	128.0	150.3	176.9	208.3	245.7	290.1	342.8	405.3
28	80.70	95.34	113.0	134.2	159.8	190.7	227.9	272.9	327.1	392.5	471.4	566.5
30	94.46	113.3	136.3	164.5	199.0	241.3	293.2	356.8	434.7	530.3	647.4	790.9

271

TABLE A-2. Future Value of Annuity (*Continued*)

n	7%	8%	9%	10%	11%	12%	13%	14%	15%	16%	17%	18%
32	110.2	134.2	164.0	201.1	247.3	304.8	376.5	465.8	577.1	715.7	888.4	1103.
34	128.3	158.6	197.0	245.5	306.8	384.5	482.9	607.5	765.4	965.3	1218.	1539.
36	148.9	187.1	236.1	299.1	380.2	484.5	618.7	791.7	1014.	1301.	1670.	2145.
38	172.6	220.3	282.6	364.0	470.5	609.8	792.2	1031.	1344.	1753.	2288.	2988.
40	199.6	259.1	337.9	442.6	581.8	767.1	1014.	1342.	1779.	2361.	3135.	4163.
45	285.7	386.5	525.9	718.9	986.6	1358.	1874.	2591.	3585.	4965.	6879.	9532.
50	406.5	573.8	815.1	1164.	1669.	2400.	3460.	4995.	7218.			
55	575.9	848.9	1260.	1881.	2818.	4236.	6380.	9623.				
60	813.5	1253.	1945.	3035.	4755.	7472.						

n	20%	22%	24%	26%	28%	30%	32%	34%	36%	38%	40%	42%
1	1.000	1.000	1.000	1.000	1.000	1.000	1.000	1.000	1.000	1.000	1.000	1.000
2	2.200	2.220	2.240	2.260	2.280	2.300	2.320	2.340	2.360	2.380	2.400	2.420
3	3.640	3.708	3.778	3.848	3.918	3.990	4.062	4.136	4.210	4.284	4.360	4.436
4	5.368	5.524	5.684	5.848	6.016	6.187	6.362	6.542	6.725	6.912	7.104	7.300
5	7.442	7.740	8.048	8.368	8.700	9.043	9.398	9.766	10.15	10.54	10.95	11.37
6	9.930	10.44	10.98	11.54	12.14	12.76	13.41	14.09	14.80	15.54	16.32	17.14
7	12.92	13.74	14.62	15.55	16.53	17.58	18.70	19.88	21.13	22.45	23.85	25.34
8	16.50	17.76	19.12	20.59	22.16	23.86	25.68	27.63	29.73	31.98	34.39	36.98
9	20.80	22.67	24.71	26.94	29.37	32.01	34.90	38.03	41.43	45.14	49.15	53.51
10	25.96	28.66	31.64	34.94	38.59	42.62	47.06	51.96	57.35	63.29	69.81	76.98
11	32.15	35.96	40.24	45.03	50.40	56.41	63.12	70.62	79.00	88.34	98.74	110.3
12	39.58	44.87	50.89	57.74	65.51	74.33	84.32	95.64	108.4	122.9	139.2	157.7
13	48.50	55.75	64.11	73.75	84.85	97.63	112.3	129.2	148.5	170.6	195.9	224.9
14	59.20	69.01	80.50	93.93	109.6	127.9	149.2	174.1	202.9	236.4	275.3	320.3
15	72.04	85.19	100.8	119.3	141.3	167.3	198.0	234.2	277.0	327.3	386.4	455.8
16	87.44	104.9	126.0	151.4	181.9	218.5	262.4	314.9	377.7	452.7	542.0	648.3
18	128.1	158.4	196.0	242.6	300.3	371.5	459.4	567.8	700.9	864.4	1065.	1310.
20	186.7	238.0	303.6	387.4	494.2	630.2	802.9	1022.	1299.	1649.	2089.	2643.
22	271.0	365.4	469.1	617.3	812.0	1067.	1401.	1837.	2405.	3142.	4097.	5332.
24	392.5	532.7	723.5	982.3	1333.	1806.	2444.	3301.	4450.	5986.	8033.	
26	567.4	795.2	1115.	1562.	2186.	3054.	4260.	5830.	8233.			
28	819.2	1186.	1716.	2482.	3583.	5164.	7426.					
30	1182.	1767.	2641.	3942.	5873.	8730.						

Note: Tables A-1, A-2, and A-3 are from *Project Economy* by Edwin S. Roscoe, Homewood, Illinois; Richard D. Irwin, Inc., 1961. Reprinted with permission.

TABLE A-3. Present Value of Annuity

Present value of annuity $[(1 + i)^n - 1]/i(1 + i)^n$ to determine present worth of a uniform series of periodic amounts. To determine periodic amount equivalent to an initial sum, use reciprocal factor (capital recovery factor).

n	.5%	1.0%	1.5%	2.0%	2.5%	3.0%	3.5%	4.0%	4.5%	5.0%	5.5%	6.0%
1	.9950	.9901	.9852	.9804	.9756	.9709	.9662	.9615	.9569	.9524	.9479	.9434
2	1.985	1.970	1.956	1.942	1.927	1.913	1.900	1.885	1.873	1.859	1.846	1.833
3	2.970	2.941	2.912	2.884	2.856	2.829	2.802	2.775	2.749	2.723	2.698	2.673
4	3.950	3.902	3.854	3.808	3.762	3.717	3.673	3.630	3.588	3.546	3.505	3.465
5	4.926	4.853	4.783	4.713	4.646	4.580	4.515	4.452	4.390	4.329	4.270	4.212
6	5.896	5.795	5.697	5.601	5.508	5.417	5.329	5.242	5.158	5.076	4.996	4.917
7	6.862	6.728	6.598	6.472	6.349	6.230	6.115	6.002	5.893	5.786	5.683	5.582
8	7.823	7.652	7.486	7.325	7.170	7.020	6.874	6.733	6.596	6.463	6.335	6.210
9	8.779	8.506	8.361	8.162	7.971	7.786	7.608	7.435	7.269	7.108	6.952	6.802
10	9.730	9.471	9.222	8.983	8.752	8.530	8.317	8.111	7.913	7.722	7.538	7.360
11	10.68	10.37	10.07	9.787	9.514	9.253	9.002	8.760	8.529	8.306	8.093	7.887
12	11.62	11.26	10.91	10.58	10.26	9.954	9.663	9.385	9.119	8.863	8.619	8.384
13	12.56	12.13	11.73	11.35	10.98	10.63	10.30	9.986	9.683	9.394	9.117	8.853
14	13.49	13.00	12.54	12.11	11.69	11.30	10.92	10.56	10.22	9.899	9.590	9.295
15	14.42	13.87	13.34	12.85	12.38	11.94	11.52	11.12	10.74	10.38	10.04	9.712
16	15.34	14.72	14.13	13.58	13.05	12.56	12.09	11.65	11.23	10.84	10.46	10.11
17	16.26	15.56	14.91	14.29	13.71	13.17	12.65	12.17	11.71	11.27	10.86	10.48
18	17.17	16.40	15.67	14.99	14.35	13.75	13.19	12.66	12.16	11.69	11.25	10.83
19	18.08	17.23	16.43	15.68	14.98	14.32	13.71	13.13	12.59	12.09	11.61	11.16
20	18.99	18.05	17.17	16.35	15.59	14.88	14.21	13.59	13.01	12.46	11.95	11.47
21	19.89	18.86	17.90	17.01	16.18	15.42	14.70	14.03	13.40	12.82	12.28	11.76
22	20.78	19.66	18.62	17.66	16.77	15.94	15.17	14.45	13.78	13.16	12.58	12.04
23	21.68	20.46	19.33	18.29	17.33	16.44	15.62	14.86	14.15	13.49	12.88	12.30
24	22.56	21.24	20.03	18.91	17.88	16.94	16.06	15.25	14.50	13.80	13.15	12.55
25	23.45	22.02	20.72	19.52	18.42	17.41	16.48	15.62	14.83	14.09	13.41	12.78
26	24.32	22.80	21.40	20.12	18.95	17.88	16.89	15.98	15.15	14.38	13.66	13.00
27	25.20	23.56	22.07	20.71	19.46	18.33	17.29	16.33	15.45	14.64	13.90	13.21
28	26.07	24.32	22.73	21.28	19.96	18.76	17.67	16.66	15.74	14.90	14.12	13.41
29	26.93	25.07	23.38	21.84	20.45	19.19	18.04	16.98	16.02	15.14	14.33	13.59
30	27.79	25.81	24.02	22.40	20.93	19.60	18.39	17.29	16.29	15.37	14.53	13.76
31	28.65	26.54	24.65	22.94	21.40	20.00	18.74	17.59	16.54	15.59	14.72	13.93
32	29.50	27.27	25.27	23.47	21.85	20.39	19.07	17.87	16.79	15.80	14.90	14.08
33	30.35	27.99	25.88	23.99	22.29	20.77	19.39	18.15	17.02	16.00	15.08	14.23
34	31.20	28.70	26.48	24.50	22.72	21.13	19.70	18.41	17.25	16.19	15.24	14.37
35	32.04	29.41	27.08	25.00	23.15	21.49	20.00	18.66	17.46	16.37	15.39	14.50
36	32.87	30.11	27.66	25.49	23.56	21.83	20.29	18.91	17.67	16.55	15.54	14.62
37	33.70	30.80	28.24	25.97	23.96	22.17	20.57	19.14	17.86	16.71	15.67	14.74
38	34.53	31.48	28.81	26.44	24.35	22.49	20.84	19.37	18.05	16.87	15.80	14.85
39	35.35	32.16	29.36	26.90	24.73	22.81	21.10	19.58	18.23	17.02	15.93	14.95
40	36.17	32.83	29.92	27.36	25.10	23.11	21.36	19.79	18.40	17.16	16.05	15.05
42	37.80	34.16	30.99	28.23	25.82	23.70	21.83	20.19	18.72	17.42	16.26	15.22
44	39.41	35.46	32.04	29.08	26.50	24.25	22.28	20.55	19.02	17.66	16.46	15.38
46	41.00	36.73	33.06	29.89	27.15	24.78	22.70	20.88	12.29	17.88	16.63	15.52

TABLE A-3.   Present Value of Annuity (*Continued*)

n	.5%	1.0%	1.5%	2.0%	2.5%	3.0%	3.5%	4.0%	4.5%	5.0%	5.5%	6.0%
48	42.58	37.97	34.04	30.67	27.77	25.27	23.09	21.20	19.54	18.08	16.79	15.65
50	44.14	39.20	35.00	31.42	28.36	25.73	23.46	21.48	19.76	18.26	16.93	15.76
52	45.69	40.39	35.93	32.14	28.92	26.17	23.80	21.75	19.97	18.42	17.06	15.86
54	47.22	41.57	36.83	32.84	29.46	26.58	24.11	21.99	20.16	18.57	17.17	15.95
56	48.74	42.72	37.71	33.50	29.96	26.97	24.41	22.22	20.33	18.70	17.28	16.03
58	50.24	43.85	38.56	34.15	30.45	27.33	24.69	22.43	20.49	18.82	17.37	16.10
60	51.73	44.96	39.38	34.76	30.91	27.68	24.94	22.62	20.64	18.93	17.45	16.16
65	55.38	47.63	41.34	36.20	31.96	28.45	25.52	23.05	20.95	19.16	17.62	16.29
70	58.94	50.17	43.15	37.50	32.90	29.12	26.00	23.39	21.20	19.34	17.75	16.38
75	62.41	52.59	44.84	38.68	33.72	29.70	26.41	23.68	21.40	19.48	17.85	16.46
80	65.80	54.89	46.41	39.74	34.45	30.20	26.75	23.92	21.57	19.60	17.93	16.51
85	69.11	57.08	47.86	40.71	35.10	30.63	27.04	24.11	21.70	19.68	17.99	16.55
90	72.33	59.16	49.21	41.59	35.67	31.00	27.28	24.27	21.80	19.75	18.03	16.58
95	75.48	61.14	50.46	42.38	36.17	31.32	27.48	24.40	21.88	19.81	18.07	16.60
100	78.54	63.03	51.62	43.10	36.61	31.60	27.66	24.50	21.95	19.85	18.10	16.62

Note: Tables A-1, A-2, and A-3 are from *Project Economy* by Edwin S. Roscoe, Homewood, Illinois; Richard D. Irwin, Inc., 1961. Reprinted with permission.

n	7%	8%	9%	10%	11%	12%	13%	14%	15%	16%	17%	18%
1	.9346	.9259	.9174	.9091	.9009	.8929	.8850	.8772	.8696	.8621	.8547	.8475
2	1.808	1.783	1.759	1.736	1.713	1.690	1.668	1.647	1.626	1.605	1.585	1.566
3	2.624	2.577	2.531	2.487	2.444	2.402	2.361	2.322	2.283	2.246	2.210	2.174
4	3.387	3.312	3.240	3.170	3.102	3.037	2.974	2.914	2.855	2.798	2.743	2.690
5	4.100	3.993	3.890	3.791	3.696	3.605	3.517	3.433	3.352	3.274	3.199	3.127
6	4.767	4.623	4.486	4.355	4.231	4.111	3.998	3.889	3.784	3.685	3.589	3.498
7	5.389	5.206	5.033	4.868	4.712	4.564	4.423	4.288	4.160	4.039	3.922	3.812
8	5.971	5.747	5.535	5.335	5.146	4.968	4.799	4.639	4.487	4.344	4.207	4.078
9	6.515	6.247	5.995	5.759	5.537	5.328	5.132	4.946	4.772	4.607	4.451	4.303
10	7.024	6.710	6.418	6.145	5.889	5.650	5.426	5.216	5.019	4.833	4.659	4.494
11	7.499	7.139	6.805	6.495	6.207	5.938	5.687	5.453	5.234	5.029	4.836	4.656
12	7.943	7.536	7.161	6.814	6.492	6.194	5.918	5.660	5.421	5.197	4.988	4.793
13	8.358	7.904	7.487	7.103	6.750	6.424	6.122	5.842	5.583	5.342	5.118	4.910
14	8.745	8.244	7.786	7.367	6.982	6.628	6.302	6.002	5.724	5.468	5.229	5.008
15	9.108	8.559	8.061	7.606	7.191	6.811	6.462	6.142	5.847	5.575	5.324	5.092
16	9.447	8.851	8.313	7.824	7.379	6.974	6.604	6.265	5.954	5.668	5.405	5.162
17	9.763	9.122	8.544	8.022	7.549	7.120	6.729	6.373	6.047	5.749	5.475	5.222
18	10.06	9.372	8.756	8.201	7.702	7.250	6.840	6.467	6.128	5.818	5.534	5.273
19	10.34	9.604	8.950	8.365	7.839	7.366	6.938	6.550	6.198	5.877	5.584	5.316
20	10.59	9.818	9.129	8.514	7.963	7.469	7.025	6.623	6.259	5.929	5.628	5.353

TABLE A-3. Present Value of Annuity (*Continued*)

n	7%	8%	9%	10%	11%	12%	13%	14%	15%	16%	17%	18%
22	11.06	10.20	9.442	8.772	8.176	7.645	7.170	6.743	6.359	6.011	5.696	5.410
24	11.47	10.53	9.707	8.985	8.348	7.784	7.283	6.835	6.434	6.073	5.746	5.451
26	11.83	10.81	9.929	9.161	8.488	7.896	7.372	6.906	6.491	6.118	5.783	5.480
28	12.14	11.05	10.12	9.307	8.602	7.984	7.441	6.961	6.534	6.152	5.810	5.502
30	12.41	11.26	10.27	9.427	8.694	8.055	7.496	7.003	6.566	6.177	5.829	5.517
32	12.65	11.43	10.41	9.526	8.769	8.112	7.538	7.035	6.591	6.196	5.844	5.528
34	12.85	11.59	10.52	9.609	8.829	8.157	7.572	7.060	6.609	6.210	5.854	5.536
36	13.04	11.72	10.61	9.677	8.879	8.192	7.598	7.079	6.623	6.220	5.862	5.541
38	13.19	11.83	10.69	9.733	8.919	8.221	7.618	7.094	6.634	6.228	5.867	5.545
40	13.33	11.92	10.76	9.779	8.951	8.244	7.634	7.105	6.642	6.233	5.871	5.548
45	13.61	12.11	10.88	9.863	9.008	8.283	7.661	7.123	6.654	6.242	5.877	5.552
50	13.80	12.23	10.96	9.915	9.042	8.304	7.675	7.133	6.661	6.246	5.880	5.554
55	13.94	12.32	11.01	9.947	9.062	8.317	7.683	7.138	6.664	6.248	5.881	5.555
60	14.04	12.38	11.05	9.967	9.074	8.324	7.687	7.140	6.665	6.249	5.882	5.555

n	20%	22%	24%	26%	28%	30%	32%	34%	36%	38%	40%	42%
1	.8333	.8197	.8065	.7937	.7812	.7692	.7576	.7463	.7353	.7246	.7143	.7042
2	1.528	1.492	1.457	1.424	1.392	1.361	1.331	1.303	1.276	1.250	1.224	1.200
3	2.106	2.042	1.981	1.923	1.868	1.816	1.766	1.719	1.673	1.630	1.589	1.549
4	2.589	2.494	2.404	2.320	2.241	2.166	2.096	2.029	1.966	1.906	1.849	1.795
5	2.991	2.864	2.745	2.635	2.532	2.436	2.345	2.260	2.181	2.106	2.035	1.969
6	3.326	3.167	3.020	2.885	2.759	2.643	2.534	2.433	2.339	2.251	2.168	2.091
7	3.605	3.416	3.242	3.083	2.937	2.802	2.677	2.562	2.455	2.355	2.263	2.176
8	3.837	3.619	3.421	3.241	3.076	2.925	2.786	2.658	2.540	2.432	2.331	2.237
9	4.031	3.786	3.566	3.366	3.184	3.019	2.868	2.730	2.603	2.487	2.379	2.280
10	4.192	3.923	3.682	3.465	2.269	3.092	2.930	2.784	2.649	2.527	2.414	2.310
11	4.327	4.035	3.776	3.543	3.335	3.147	2.978	2.824	2.683	2.555	2.438	2.331
12	4.439	4.127	3.851	3.606	3.387	1.190	3.013	2.853	2.708	2.576	2.456	2.346
13	4.533	4.303	3.912	3.656	3.427	3.223	3.040	2.876	2.727	2.592	2.469	2.356
14	4.611	4.265	3.962	3.695	3.459	3.249	3.061	2.892	2.740	2.603	2.478	2.363
15	4.675	4.315	4.001	3.726	3.483	3.268	3.076	2.905	2.750	2.611	2.484	2.369
16	4.730	4.357	4.033	3.751	3.503	3.283	3.088	2.914	2.757	2.616	2.489	2.372
18	4.812	4.419	4.080	3.786	3.529	3.304	3.104	2.926	2.767	2.624	2.494	2.377
20	4.870	4.460	4.110	3.808	3.546	3.316	3.113	2.933	2.772	2.627	2.497	2.379
22	4.909	4.488	4.130	3.822	3.556	3.323	3.118	2.936	2.775	2.629	2.498	2.380
24	4.937	4.507	4.143	3.831	3.562	3.327	3.121	2.939	2.776	2.630	2.499	2.380
26	4.956	4.520	4.151	3.837	3.566	3.330	3.123	2.940	2.777	2.631	2.500	2.381
28	4.970	4.528	4.157	3.840	3.568	3.331	3.124	2.940	2.777	2.631	2.500	2.381
30	4.979	4.534	4.160	3.842	3.569	3.332	3.124	2.941	2.778	2.631	2.500	2.381

Note: Tables A-1, A-2, and A-3 are from *Project Economy* by Edwin S. Roscoe, Homewood, Illinois; Richard D. Irwin, Inc., 1961. Reprinted with permission.

References

1. Baeck, Henry S., *Practical Servomechanism Design.* McGraw-Hill Book Company, N.Y., 1968.
2. Grabbe, Eugene M., Ramo, Simon, and Wooldridge, Dean E., *Handbook of Automation, Computation, and Control.* John Wiley & Sons, N.Y., 1958.
3. DiStefano, Joseph J., III, Stubberud, Allen R., and Williams, Ivan J., *Feedback and Control Systems.* Schaum's Outline Series. McGraw-Hill Book Company, N.Y., 1967.
4. *Professional Engineering Examinations, Volume I.* National Council of Engineering Examiners, 1972.
5. *Solutions: Professional Engineering Examinations, Volume II, 1965–1971.* National Council of Engineering Examiners, 1974.
6. Stout, Melville B., *Basic Electrical Measurements.* Prentice-Hall, Inc., N.Y., 1955.
7. Timbie, William H., Bush, Vannevar, and Hoadley, George B., *Principles of Electrical Engineering.* John Wiley & Sons, N.Y., 1953.
8. Kerchner, Russell M., and Corcoran, George F., *Alternating-Current Circuits.* John Wiley & Sons, N.Y., 1955.
9. Lawrence, Ralph R. and Richards, Henry E., *Principles of Alternating-Current Machinery.* McGraw-Hill Book Company, Inc., N.Y., 1953.
10. Kloeffler, Royce G., Kerchner, Russell M., and Brenneman, Jesse L., *Direct-Current Machinery.* The Macmillan Company, N.Y., 1955.
11. Gardner, Murray F. and Barnes, John L., *Transients in Linear Systems.* John Wiley & Sons, Inc., N.Y., 1957.
12. Reed, Henry R. and Corcoran, George F., *Electrical Engineering Experiments.* John Wiley & Sons, Inc., N.Y., 1950.
13. Fich, Sylvan, *Transient Analysis in Electrical Engineering.* Prentice Hall, Inc., N.Y., 1955.
14. Kurtz, Edwin B. and Corcoran, George F., *Introduction to Electric Transients.* John Wiley & Sons, Inc., N.Y., 1953.
15. Skilling, Hugh Hildreth, *Fundamentals of Electric Waves.* John Wiley & Sons, Inc., N.Y., 1956.
16. Kraus, John D., *Electromagnetics.* McGraw-Hill Book Company, Inc., N.Y., 1953.
17. Ryder, John D., *Networks, Lines and Fields.* Prentice Hall, Inc., N.Y., 1955.
18. Ramo, Simon, and Whinnery, John R., *Fields and Waves in Modern Radio.* John Wiley & Sons, Inc., N.Y., 1953.
19. Martin, Thomas L., Jr., *Electronic Circuits.* Prentice Hall, Inc., Englewood Cliffs, N.J., 1956.
20. Ryder, John D., *Electronic Engineering Principles.* Prentice Hall, Inc., Englewood Cliffs, N.J., 1955.
21. Shea, Richard F., *Transistor Circuit Engineering.* John Wiley & Sons, Inc., N.Y., 1957.
22. Cutler, Phillip, *Semiconductor Circuit Analysis.* McGraw-Hill Book Company, N.Y., 1964.
23. Lenk, John D., *Handbook of Simplified Solid-State Circuit Design.* Prentice Hall, Inc., Englewood Cliffs, N.J., 1971.
24. Rainville, Earl D., *Elementary Differential Equations.* The Macmillan Company, N.Y., 1955.
25. Phister, Montgomery, Jr., *Logic Design of Digital Computers.* John Wiley & Sons, Inc., N.Y., 1958.
26. Hill, Fredrick J. and Peterson, Gerald R., *Introduction to Switching Theory and Logical Design.* John Wiley & Sons, Inc., N.Y., 1968.
27. Milsum, John H., *Biological Control Systems Analysis.* McGraw-Hill Book Company, N.Y., 1966.
28. Park, William R., *Cost Engineering Analysis.* John Wiley & Sons, Inc., N.Y., 1973.

REFERENCES

29. Kurtz, Max, *Engineering Economics for Professional Engineer's Examinations.* McGraw-Hill Book Company, N.Y., 1975.
30. Graeme, Jerald G., *Applications of Operational Amplifiers, Third Generation Techniques.* McGraw-Hill Book Company, N.Y., 1973.
31. *CRC Standard Mathematical Tables.* The Chemical Rubber Company, Cleveland, Ohio, 1973.
32. Jones, Lincoln D., *Electrical Engineering License Review.* Engineering Press, San Jose, California, 1972.
33. Lyons, John S. and Dublin, Stanley W., *Electrical Engineering and Economics and Ethics for Professional Engineering Examinations.* Hayden Book Company, Inc., N.Y., 1970.
34. *DC Motors, Speed Controls, Servo Systems.* Electro-Craft Corporation, Hopkins, Minnesota, 1973.
35. Souders, Mott, *The Engineer's Companion.* John Wiley & Sons, Inc., N.Y., 1967.
36. *Reference Data for Radio Engineers.* Howard Sams and Company, Inc., Indianapolis, Indiana, 1968.
37. Baumeister, Theodore, *Marks' Standard Handbook for Mechanical Engineers.* McGraw-Hill Book Company, N.Y., 1967.
38. Langford-Smith, F., *Radiotron Designer's Handbook.* RCA, Harrison, N.J., 1954.
39. Knowlton, A. E., *Standard Handbook for Electrical Engineers.* McGraw-Hill Book Company, Inc., N.Y., 1949.
40. *An Applications Guide for Operational Amplifiers.* Application Note AN-20, National Semiconductor Corporation, Santa Clara, California, February, 1969.
41. Dobkin, Robert C., *Op Amp Circuit Collection.* Application Note AN-31, National Semiconductor Corporation, Santa Clara, California, February, 1970.
42. *FET Circuit Applications.* Application Note AN-32, National Semiconductor Corporation, February, 1970.
43. Hunt, Pearson, Williams, Charles M., and Donaldson, Gordon, *Basic Business Finance.* Richard D. Irwin, Inc., Homewood, Illinois, 1961.
44. Hill, Thomas M. and Gordon, Myron J., *Accounting: A Management Approach.* Richard D. Irwin, Inc., Homewood, Illinois, 1959.
45. Johnson, Robert W., *Financial Management.* Allyn and Bacon, Inc., Boston, 1962.
46. Lewis, John P., *Business Conditions Analysis.* McGraw-Hill Book Company, Inc., N.Y., 1959.
47. *National Electric Code 1975*, National Fire Protection Association, Boston, MA.
48. Bentley, James H., "The Foolproof Way to Sequencer Design," *Electronic Design* 10, (May 10, 1973).
49. Millman, Jacob and Halkias, Christos C., *Integrated Electronics: Analog and Digital Circuits and Systems*, McGraw-Hill Book Company, Inc., N.Y., 1972.
50. Summers, Wilford I., and Watt, John H., *NFPA Handbook of the National Electric Code*, 4th Edition, McGraw-Hill Book Company, N.Y., 1975.
51. Illuminating Engineering Society, *IES Lighting Handbook*, 5th Edition, N.Y., 1972.
52. Baldwin, Allen J., and Hess, Karen M., *A Programmed Review of Engineering Fundamentals*, Van Nostrand Reinhold Company, N.Y., 1978.
53. *Professional Engineering Examinations, Volume III, 1972–1976*, National Council of Engineering Examiners, 1979.

Index

INDEX

INDEX

INDEX